Building Green: Exploring Bamboo as a Sustainable Construction Material

Purky

TABLE OF CONTENTS

Chapter 1

1   INTRODUCTION

1.1   Background

The Earth's atmosphere is composed of oxygen, nitrogen, water vapour and greenhouse gasses (GHG), such as carbon dioxide (CO_2), methane (CH_4) and nitrous oxide (N_2O), among others. When the Earth is heated by the sun, infrared heat is emitted, some of which is trapped in the atmosphere by the GHGs. This heating process results in the Earth's surface being, on average, approximately 15°C (IPCC, 2018). If the concentration of GHGs in the atmosphere increases, more heat is trapped, and the Earth's surface becomes warmer, a phenomenon known as global warming.

Over a 100-year period from 1906 – 2005, the average global temperature rose by 0.74°C, mainly due to the increased emission of GHGs from human activity (CSIR, 2015). In the intervening years, the global temperature has risen further to 1.0°C, following the warming trend, as indicated in Figure 1.1. It is predicted that if these GHG emissions continue at the current rates, global warming will reach 1.5°C above pre-industrial temperatures by approximately the year 2040 (IPCC, 2018).

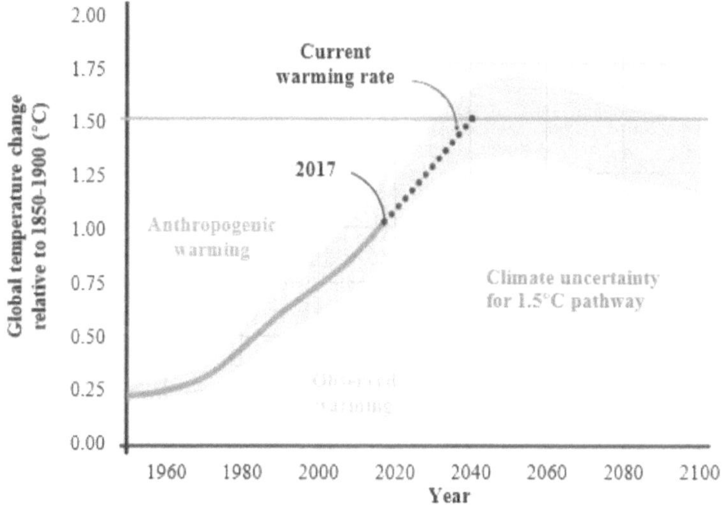

Figure 1.1: Anthropogenic (human) warming of the Earth's surface (IPCC, 2018)

If global warming continues, this could represent a potentially irreversible threat to human societies and the Earth. However, even if in the future no further GHGs are emitted, the global temperature will remain high for centuries, due to the slow removal of carbon dioxide from the atmosphere. Therefore, at the 21st conference of the United Nations Framework Convention on Climate Change

(UNFCCC) in December 2015, 195 nations adopted the Paris Agreement, with the aim of restricting the temperature increase to a maximum of 1.5°C above pre-industrial levels (IPCC, 2018).

In order to stabilise the global temperature, it would be necessary to stop the increase of GHGs, specifically carbon dioxide, in the atmosphere. This can be done in two ways – either by reducing the amount of GHG emissions or by removing carbon dioxide from the atmosphere.

In 2000, 65% of GHG emissions were from the power, transport, buildings, and industry sectors (refer Figure 1.2). The emissions grouped under 'Energy Emissions' were mostly CO_2 emissions. Just under a quarter of the GHG emissions were from the generation of power and heat, from domestic, commercial, and industrial use. This sector also included emissions from coal mines, gas works and petrol refineries using fossil fuels.

The emissions reported under the transport sector included emissions from road transport, aviation, rail and shipping, while the building sector accounted for the direct combustion of fossil fuels and biomass in residential and commercial buildings, mostly for cooking and heating.

Industry accounted for 14% of the GHG emissions, of which 10% were from manufacturing and construction, and 4% from industrial processes such as the production of chemicals and cement (Stern, 2007).

Thus, in 2000, the construction industry could be considered responsible for 10% of the global GHG emissions.

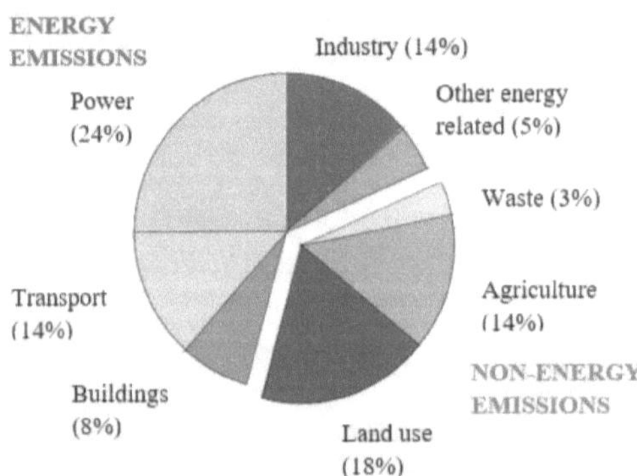

Figure 1.2: Global GHG emissions by sector in 2000 (Stern, 2007)

By 2014, it was reported that building construction had decreased to 9% of CO_2 emissions (refer Figure 1.3), indicating a decrease of 2% (IEA, 2016). These emissions increased again to 11% in 2018 (refer Figure 1.4) (IEA, 2019).

(The 8% and 28% CO_2 emissions attributed to 'Buildings' in Figure 1.2, Figure 1.3 and Figure 1.4

refer to the emissions generated by buildings during their lifetime, mainly for heating and cooling, also referred to as operational carbon emissions)

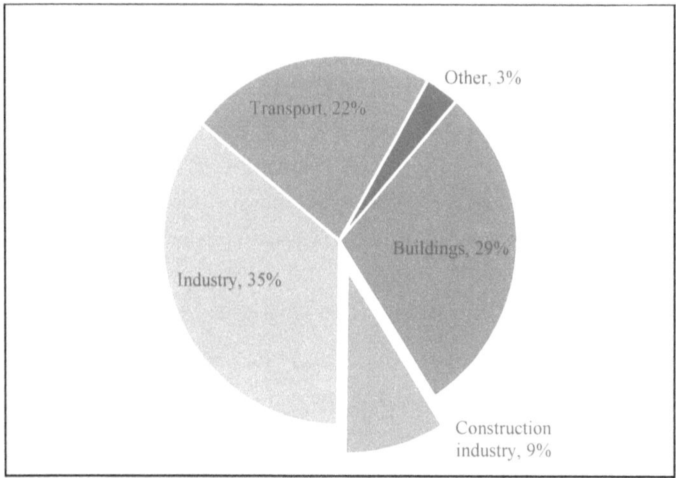

Figure 1.3: Global CO₂ emissions by sector in 2014 (IEA, 2016)

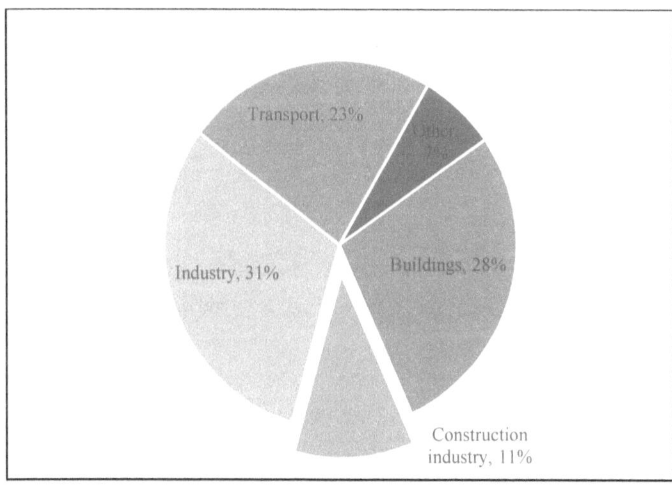

Figure 1.4: Global CO₂ emissions by sector in 2018 (IEA, 2019)

It should be noted that reports produced by different bodies or groups often provide conflicting numbers regarding GHG emissions. The percentages presented here were extracted from global reports that focussed explicitly on the building and construction industry, as these reports were deemed to have the most accurate figures for this topic.

Buildings contribute to carbon emissions in how they are constructed (embodied carbon emissions) and how they are used (operational carbon emissions) (Budds, 2019), (Becqué, et al., 2019). Structural and Materials Engineers have minimal input on the operational use of buildings but can influence the construction methods as well as the materials used. In this way, they can affect the extent of carbon emissions during the construction of a building.

In order to determine the carbon emissions of a material or building product, the carbon footprint is measured, which is a methodology to measure the GHG emission during its life cycle (van der Lugt & Vogtlander, 2015). In recent times, the term embodied carbon of construction materials and products has come to be synonymous with the term carbon footprint (Sansom & Pope, 2012), where "embodied carbon" is the total amount of carbon dioxide (CO_2) emitted by a material throughout its life cycle (Kaminski, et al., 2016). This is covered in more detail in Chapter 2.

Many residential, commercial and industrial buildings globally and in South Africa are built with the traditional building technology choices of structural steelwork, reinforced concrete, and clay or concrete masonry. These have been reported as contributing to the GHG emission problem (Kibert, 2016), (Muigai, 2014). Although there is a shift towards using more sustainable and "green" materials, the clients' preference remains directed to using "tried-and-tested" building technologies, particularly from governmental departments. This could be due to lack of knowledge, concern over construction or maintenance issues, or cost implications.

Although there is a desire to build more sustainably, the change to alternative building technologies is hampered by the lack of cogent and coherent material and structural information to support appropriate choices of building material.

1.1.1 Bamboo construction as a sustainable building technology

Bamboo has been used as a traditional building material for shelters for centuries, with reports of bamboo houses in China from 2000 years ago (Wang & Guo, 2001), and dams such as the Du Jiang Dam built during the Xia Dynasty (Krawczuk, 2013). In recent times, bamboo has been used to build cathedrals in Colombia (refer Figure 1.5), schools such as the Green School in Indonesia (refer Figure 1.6) and bridges such as the bamboo bridge in China (refer Figure 1.7) (Krawczuk, 2013).

Research has shown the suitability of this material in terms of structural strength for various species of bamboo, and many architects have praised its aesthetic qualities. However, in most parts of the world, bamboo is seen to have connotations with poverty. There have been many attempts to address this paradox by scientists, engineers and architects, through projects such as the National Bamboo Project in Costa Rica, and research by organisations such as INBAR (International Network for Bamboo and Rattan). Furthermore, bamboo plants have been touted as preferable building materials in terms of sustainability and ecological impact, and this book aims to explore this claim based on available literature studies.

Figure 1.5: Bamboo cathedral by Simón Vélez, Pereira, Colombia (Krawczuk, 2013)

Figure 1.6: Bamboo Green School, Bali, Indonesia (Krawczuk, 2013)

Figure 1.7: Bamboo bridge by Simón Vélez, China (Krawczuk, 2013)

1.2   Research Aims and Objectives

This book aims to contribute towards the use and understanding of alternative and sustainable building materials and technologies by collating information on the use of whole culm bamboo as a structural material, as well as the recommended structural design procedures to be followed with this material.

Based on this information, the book aims to evaluate whether bamboo construction is a viable construction technology for South Africa.

In order to achieve this, the following objectives need to be attained:

1) To collate and evaluate available literature regarding the material and engineering properties of various species of bamboo.

2) To collate and critique the proposed design approaches to designing with whole bamboo culms, including the availability of design codes and guidelines.

3) To present case studies to demonstrate the variety of structures that can be built with bamboo culms.

4) To collate information regarding the current status of bamboo availability and use in South Africa, and to assess the viability of bamboo construction in South Africa.

5) To identify aspects of whole culm bamboo construction that require further research and testing.

1.3   Key Research Questions/Problem Definition

The preliminary literature review indicated that there was a lack of scientific or structural information regarding bamboo as a building material from an engineering or materials perspective, both globally as well as in South Africa. A large portion of the information was provided by architects, who promoted bamboo as a favourable alternative to timber or steel from a sustainability perspective. However, few scientific facts were published to corroborate these statements.

Furthermore, due to the many different species of bamboo, and the vagaries of working with a natural material, i.e. not man-made, the qualitative information found regarding the material and engineering properties varied widely, and it was challenging to establish engineering design parameters.

Furthermore, the information found relating to bamboo in South Africa focussed mainly on bamboo as a source for biofuel, and the growing of bamboo for such purposes.

These issues made it clear that if bamboo construction were to become an accepted form of construction in South Africa, a fair amount of research was still needed. This book will attempt to summarise the critical information currently available in order to highlight areas requiring further research and investigation in the South African context.

The key questions were:

- What can the building and construction sector do to reduce Greenhouse Gas emissions?
- What is bamboo, and where is it found?
- Can bamboo be considered a sustainable construction material?
- What are the engineering and material properties of bamboo?
- What are the existing standards and design codes for bamboo construction?
- What design procedures would be used to design a bamboo structure?
- Are there examples of engineered bamboo structures?
- Are there any limitations to designing with bamboo?
- Is bamboo construction a viable building technology for South Africa?

1.4   Research Methodology

As this book is, to a large extent, a collation and evaluation of existing information on bamboo construction, various approaches were used to source the information.

1.4.1   Literature review and document analysis

A substantial and comprehensive literature review and document analysis were conducted on the technical and environmental aspects of bamboo construction, both globally and locally in South Africa. Due to the limited technical data available in written form in South Africa, information was gathered mostly from online news articles, project reports and internet websites. Scholarly articles of bamboo in various countries, mainly covering technical or engineering aspects of bamboo, were sourced online and reviewed. Structural journals were also examined for reports or papers pertaining to bamboo construction.

Each document or article retrieved was critiqued for the reliability and validity of the information presented, by comparison with similar articles. Also, articles or documents that submitted statements regarding the properties of bamboo plants without numerical statistics were discarded.

1.4.2   Site visits and Case Studies

As the existing bamboo structures in South Africa were extremely limited, the only site visits were to local areas of bamboo growth and a bamboo farm in Vredendal, in the Western Cape province of South Africa.

Case studies were selected that showcased the variety of structures that could be built from bamboo. The architects and builders of each case study were contacted via email for information regarding the various structures, and feedback incorporated where available. The majority of the feedback consisted of additional published documents and photographs.

1.5 Scope and Limitations

1.5.1 Scope of the book

The book focusses on the use of whole culm bamboo used in buildings and structures, such as bridges or residential houses. It does not cover temporary structures such as scaffolding, nor does it cover bamboo products, such as flooring boards, in detail.

The book is a review of information available in the literature, as well as in practice.

1.5.2 Selected parameters

Due to the vast number of bamboo species in existence worldwide, the detailed review of the various engineering and material properties of all species was not possible within the scope of this book. The International Network for Bamboo and Rattan (INBAR) identified five main species as suitable for structural purposes, namely *Bambusa balcooa, Bambusa nutans, Bambusa polymorpha, Gigantochloa apus* and *Guadua angustifolia* (Jayanetti & Follett, 1998). These were selected for more detailed review. Another six species commonly used in construction were also included, namely *Bambusa vulgaris, Phyllostachys pubescens, Bambusa stenostachya, Dendrocalamus strictus, Dendrocalamus asper* and *Dendrocalamus giganteus* (Richard, 2013).

1.5.3 Limitations

The lack of bamboo construction in South Africa was not considered a limiting condition, as this was one of the reasons for this book.

However, this lack of South African examples resulted in the selection of case studies from different countries, which meant that communication with the designers was often limited. As non-engineers often drove the case studies, it was problematic to extract details from an engineering perspective.

The vast amount of information available on bamboo on the Internet was problematic, in that much of it was irrelevant or written without engineering substantiation. This necessitated the reviewing of multiple documents to select the relevant few that covered the information sought for this book.

The book is a review of information available in the literature, as well as in practice. No laboratory tests were performed on any bamboo samples, and simple design calculations were done only for illustration purposes.

1.6 Outline of book

The book is broken up into the following sections:

Chapter 1 presents the main objective of the book, namely, to collate information on the use of bamboo as a sustainable structural material, and to evaluate whether bamboo construction is a viable construction technology for South Africa.

Chapter 2 provides a brief background to Sustainability, Sustainable Construction and alternative building technologies, and why bamboo can be considered an environmentally suitable building material in terms of Sustainable Construction.

Chapter 3 describes the anatomical and physiological aspects of the bamboo plant and identifies
species to be considered in greater detail throughout the remainder of the book.

Chapter 4 discusses the properties of bamboo species considered suitable for structural construction
applications, with a brief mention of bamboo building products.

Chapter 5 considers current design codes, standards and guidelines for bamboo construction, and
discusses design approaches for the various stress and building systems.

Chapter 6 presents four case studies from relevant global examples to illustrate the viability and
possibilities available for the use of full-culm bamboo in buildings and structures in residential,
industrial, commercial and transportation sectors.

Chapter 7 provides a summary of the history of the bamboo industry in South Africa and evaluates
the use of bamboo construction in South Africa, based on bamboo currently grown in South Africa.

Chapter 8 concludes the book, highlighting areas that require further research for bamboo construction to become a viable building technology in South Africa with locally grown species.

1.7   References

Anon., 2016. *Building Code.* [Online] Available at: http://en.wikipedia.org/wiki/Building_code [Accessed 20 January 2016].

Becqué, R. et al., 2019. *Accelerating Building Decarbonization: Eight Attainable Policy Pathways to Net Zero Carbon Buildings for All,* Washington, DC: World Resources Institute.

Budds, D., 2019. *How do buildings contribute to climate change?* [Online] Available at: www.curbed.com/2019/9/19/20874234/buildings-carbon-emissions-climate-change [Accessed 20 January 2020].

Bureau of Indian Standards, 2016. *National Building Code of India,* s.l.: Bureau of Indian Standards.

Canavan, S. et al., 2017. The global distribution of bamboos: assessing correlates of introduction and invasion. *AoB Plants,* Volume 9.

Cheremisinoff, N. P. & Rosenfeld, P. E. eds., 2010. *Handbook of Pollution Prevention and Cleaner Production.* s.l.: William Andrew Publishing.

CSIR, 2015. *Potential for reducing Greenhouse Gas Emissions in the South African Construction Sector,* Pretoria: CSIR.

Gottron, J. M., 2014. *Structural Performance of Bamboo: Capacity under sustained loads and monotonic bending,* Pittsburgh: University of Pittsburgh.

IBC, 2015. *International Building Code.* USA: International Code Council.

ICE, 2019. *Embodied Carbon - The ICE Database.* [Online] Available at: http://www.circularecology.com/ice-database.html

IEA, 2016. *2016 Global Status Report: Towards zero-emission efficient and resilient buildings,* s.l.: Global Alliance for Buildings and Construction.

IEA, 2019. *2019 Global Status Report for Buildings and Construction: Towards a zero-emission, efficient and resilient buildings and construction sector,* s.l.: Global Alliance for Buildings and Construction.

Illsley, C., 2018. *What are Native, Indigenous and Endemic Species?* [Online] Available at: https://www.worldatlas.com/articles/what-is-the-difference-between-a-native-an-indigenous-and-an-endemic-species.html [Accessed 12 June 2019].

IPCC, 2018. *Global Warming of 1.5°C,* s.l.: IPCC.

Janssen, J. J. A., 2000. *Designing and building with bamboo,* Beijing: INBAR.

Jayanetti, L. & Follett, P., 1998. *Technical Report 16: Bamboo in Construction - An Introduction,* Beijing: INBAR.

Kaminski, S., Lawrence, A. & Trujillo, D., 2016. *Design Guide for Engineered Bahareque Housing,* Beijing: INBAR.

Kibert, C., 2016. *Sustainable Construction - Green Building Design and Delivery.* 4th ed. Hoboken, New Jersey: John Wiley & Sons, Inc.

Kibert, C. J., Sendzimir, J. & Guy, G. B., 2002. *Construction Ecology - Nature as the basis for green buildings.* London: Spoon Press.

Krawczuk, K., 2013. *Bamboo as a sustainable material for future building industry,* Københavns: Københavns Erhvervsakademi.

Mena, J., Vera, S., Correal, J. F. & Lopez, M., 2012. Assessment of fire reaction and fire resistance of Guadua angustifolia kunth bamboo. *Construction and Building Materials,* Issue 27, pp. 60-65.

Muigai, R. N., 2014. *A framework towards the design of more sustainable concrete structures,* Cape Town: University of Cape Town.

Richard, M. J., 2013. *Assessing the Performance of Bamboo Structural Components,* Pittsburgh: University of Pittsburgh.

RICS, 2010. *Global Glossary of Sustainability Terms.* [Online] Available at: http://www.trentglobal.edu.sg/wp-content/uploads/2016/09/RICS-Sustainability-Glossary-June-2010.pdf [Accessed 2015].

Sansom, M. & Pope, R., 2012. A comparative embodied carbon assessment of commercial buildings. *The Structural Engineer,* October.pp. 38-49.

Scheba, A., Blanchard, R. & Mayeki, S., 2017. *Bamboo for green development? The opportunities and challenges of commercialising bamboo in South Africa,* Cape Town: HSRC.

Stern, N., 2007. *Stern Review: The Economics of Climate Change,* Cambridge: Cambridge University Press.

Stubbs, B., 2008. *Plain English Guide to Sustainable Construction,* London: Constructing Excellence.

UN Environment and International Energy Agency, 2017. *Global Status Report 2017,* s.l.: UN.

van der Lugt, P. & Vogtländer, J. G., 2015. *The Environmental Impact of Industrial Bamboo Products: Life-cycle Assessment and Carbon Sequestration,* Beijing: INBAR.

Wang, Z. & Guo, W., 2001. *Current status and prospects of new architectural materials from bamboo,* Beijing: INBAR.

WCED (World Commission on Environment and Development), 1987. *Our Common Future*, Oxford: Oxford University Press.

Winter Park Bamboo Inc., 2015. *Bamboo Glossary*. [Online] Available at: www.winterparkbamboo.com/bamboo-glossary/ [Accessed 4 November 2017].

Chapter 2

2 BACKGROUND TO SUSTAINABILITY AND ALTERNATIVE BUILDING TECHNOLOGIES

2.1 Introduction

The concept of sustainability is not new. The first document articulating the need for engineers to design sustainably was produced in 1925 by Professor Svante August Arrhenius, the Director of the Nobel Institute in Sweden at the time. In his work, *Chemistry in Modern Life*, he suggested that engineers design engines capable of operating with alternative fuels and suggested that wind and solar energy would reduce the level of CO_2 emissions. He also stated that the amount of industry waste should be reduced, to ensure that future generations could also meet their living needs (Smith, et al., 2007).

However, the term 'sustainability' first became part of the environmental terminology in 1987 with the Brundtland Report, which described sustainability as "leaving sufficient resources for future generations to have a quality of life similar to ours" (Kibert, 1994). This report led to an environmental movement which gradually included other segments of society, including construction.

Traditionally, the principal objective in structural building projects is to produce a satisfactory end product with the least cost in the shortest period of time (refer Table 2.1). This approach often results in the increased use of fast-track technologies, such as precast concrete members and structural steelwork frames, which typically use materials with a high embodied energy (Singh, 2007).

In contrast, the main objective of sustainability is to protect the Earth's resources while maintaining a healthy living environment for all. The challenge for the construction industry is how to meet the sustainability criteria while maintaining the traditional criteria of high performance and quality, coupled with an acceptable cost.

Table 2.1: Traditional and Sustainability criteria for building materials and products (Kibert, 1994)

Traditional Criteria	Sustainability Criteria
• Performance • Quality • Cost	• Resource depletion • Environmental degradation • Healthy environment

2.2 Sustainability, Sustainable Development and Sustainable Construction

Buildings have not always been built unsustainably. The last few decades have seen improvements in operational energy efficiency, but also an increase in the embodied energy of building materials. At the same time, in the last 100 years, the average house size has doubled for the same number of occupants (Landman, 1999). Prior to air conditioning and fluorescent lighting, buildings were designed with better ventilation and daylighting than current standard practice.

At the time of the Industrial Revolution (1760 - 1840), now also known as the First Industrial Revolution, buildings were built for the workers, which did not allow for adequate air and light. In the early 20th century, attempts were made to improve these living conditions. Air and sunlight were considered a cure for many health ailments prevalent at that time, such as consumption and tuberculosis.

In the early 1900s, the "garden cities" movement began in Europe. Solar orientation was considered in these designs for health reasons, and the "garden cities" concept suggested the use of local construction materials.

At the Stockholm Conference on the Human Environment in 1972, it was first acknowledged that the goals for economic growth and development in the Western world conflicted with the protection of the natural environment (Kates, et al., 2005). At the same time, the Club of Rome commissioned the "Limits to Growth" report, which described the consequences of an increasing global population in a world with finite resources.

In 1980, at the World Conservation Strategy of the International Union for the Conservation of Nature, the concept of environmental conservation was argued as a means of achieving sustainability (Mebratu, 1998).

Shortly after this, in 1987, the World Commission on Environment and Development (WCED) published "Our Common Future", a document highlighting environmental concerns and bringing the concepts of sustainability and sustainable development to the attention of the construction industry. Gro Brundtland, then prime minister of Norway, chaired the WCED and thus the publication became known as the Brundtland Commission. This publication is often seen as the starting point for sustainable development, with the most common definition for sustainable development being "development that meets the needs of the present without compromising the ability of future generations to meet their own needs" (Bruntlandt, 1987).

In 1992, the United Nations Conference on Environment and Development (UNCED), also known as the Earth Summit, was held in Rio de Janeiro, Brazil. A set of principles for sustainable development was developed, known as Agenda 21, and signed by 178 countries, including South Africa. This document was considered a blueprint for sustainable development and reflected a global commitment to integrating environmental concerns into social and economic decision-making processes.

A further outcome of the Earth Summit was the introduction of the United Nations Framework Convention on Climate Change (UNFCCC), which was established to address the impacts of climate change, effective from March 1994, after it had been ratified by several countries. The UNFCCC objective was for industrialised countries to stabilise greenhouse gas emissions.

The Kyoto Protocol was written in 1992 as an extension to the UNFCCC, establishing legally binding obligations for the countries to reduce their GHG emissions in the period 2008-2012. In 2012 the Protocol was amended to include the period 2013-2020, referred to as the Doha Amendment. The Paris Agreement was subsequently adopted in 2015, governing emission reductions from 2020 onwards, with the aim of limiting GHG emissions to no more than 1.5°C above pre-industrial levels (IPCC, 2018).

Following this, several conferences were held on sustainability issues, such as the United Nations Conference, Habitat II, held in Istanbul in 1996, and the United Nations Earth Summit II in 1997, where the progress and implementation of Agenda 21 were assessed (The Institute of Structural Engineers, 1999).

In 2002 at the World Summit on Sustainable Development (WSSD) held in Johannesburg, South Africa, the commitment to sustainable development and the objectives and goals set out in Agenda 21 were re-affirmed.

Subsequently, the United Nations Conference on Sustainable Development (Rio + 20) was held in 2012, 20 years after the Earth Summit held in Rio de Janeiro in 1992. One outcome of this Conference was a 10-year framework of programmes on sustainable consumption and production patterns for the period from 2012 to 2022. These programmes included protecting and restoring the Earth's ecosystems, reducing the use of hazardous materials and toxic chemicals, the reduction of emission of pollutants, protecting natural resources and promoting more efficient usage of natural resources and products (Leggett & Carter, 2012).

In order to understand the importance of the conferences and publications as mentioned above, it is necessary to understand the terms used. This is dealt with in more details in the following section.

2.2.1   Definitions: Sustainability, Sustainable Development and Sustainable Construction

The term 'sustainability' was originally defined in the field of ecology, referring to an ecosystem's ability to exist over time with little or no change (Jabareen, 2008). This term later came to mean "the existence of the ecological conditions necessary to support human life at a specified level of well-being through future generations", called 'ecological sustainability' by Lelé (Lélé, 1991). Adding the concept of 'development' shifted the focus from environment to society. This shift can be seen in the definition of Sustainable Development by the World Commission on Environment and Development (WCED), which defines sustainable development as "development that meets the needs of the present without compromising the ability of future generations to meet their own needs" (Brundtland, 1987). This is often considered the starting point for the concept of sustainable development.

The charter of the World Business Council for Sustainable Development (WBCSD) elaborates on this further. It states that the concept of sustainable development "recognises that economic growth and environmental protection are inextricably linked and that the quality of present and future life rests on meeting basic human needs without destroying the environment upon which all life depends" (Mebratu, 1998).

Both the charter of the WBCSD and the Bruntlandt report have three main goals, namely environmental, economic and social sustainability. These three aspects of sustainable development are referred to as the three pillars of sustainability and are often represented as interlocking rings (refer Figure 2.1). These indicate that the different areas are independent of each other and equally important, but with areas that overlap. The overlapping area of the three rings represents the integration area for sustainable development (Mebratu, 1998).

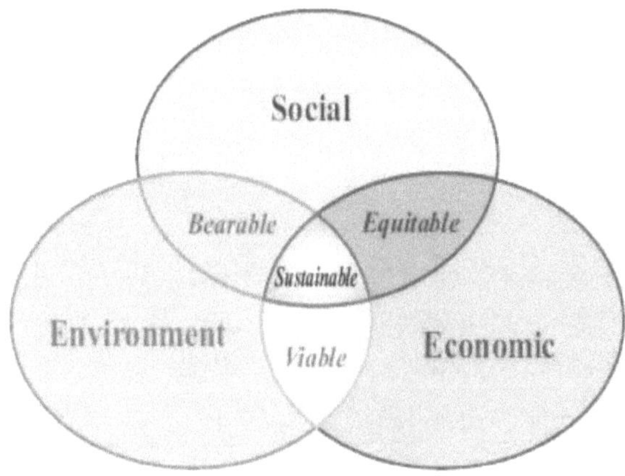

Figure 2.1 Interlocking ring model for sustainable development (Stubbs, 2008)

Various definitions can be found in the literature regarding "sustainable construction" or "sustainable building", terms often used interchangeably with "green building". Kibert defines *sustainable building and construction* as "creating a healthy built environment based on ecologically sound principles" (Kibert, et al., 2002). According to the Green Building Council of Australia, the concept of a *green building* addresses areas such as greenhouse gas emissions, energy efficiency, water conservation, waste avoidance, pollution prevention, reuse and recycling, reduced resource consumption, enhanced biodiversity and productive, healthy environments (Hayles & Holdsworth, 2008), thus applying the principles of *sustainable construction* to buildings.

The Royal Institution of Chartered Surveyors (RICS) defines a green building as "an outcome of a design philosophy which focuses on increasing the efficiency of resource use, including energy, water and materials, while reducing building impacts on human health and the environment during the building life cycle through better siting, design, construction, operation, maintenance and removal" (RICS,

2010).

According to Singh (2007), sustainable construction can be defined as a suitable choice of materials, sources, construction methodologies and design philosophies to increase performance, reduce environmental impact and waste, and to be less harmful to the environment.

Valdes et al. (2018) produced a diagram to show how the concepts of Sustainable Construction and Sustainable Development are linked (refer Figure 2.2). In this diagram, Valdes highlights seven principles of sustainable construction as proposed by Kibert (2016), namely reduce resource consumption, reuse resources, recycle, protect nature, eliminate toxic materials or substances, focussing on quality and the application of life-cycle costing.

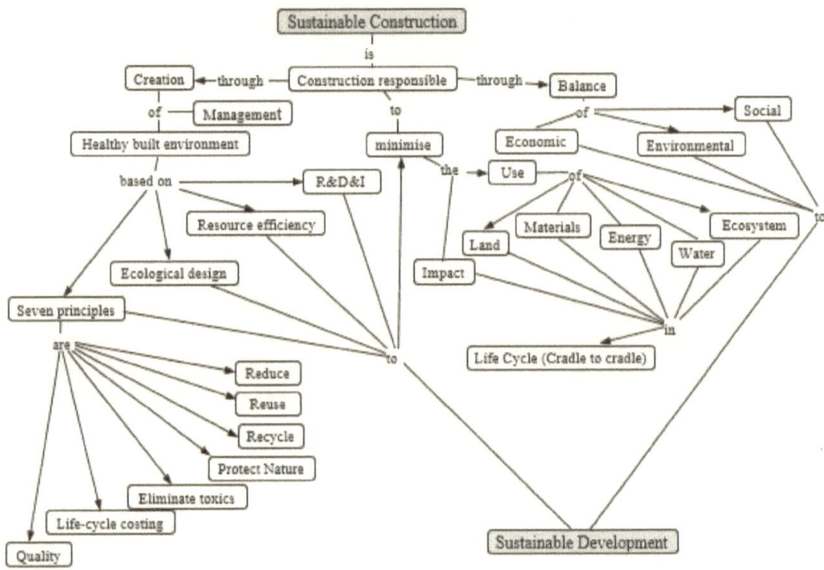

Figure 2.2: Conceptual map of Sustainable Construction and Sustainable Development (Valdes, et al., 2018)

2.2.2   Climate Change - Globally and in South Africa

Over the past four decades, there has been a global shift towards protecting the environment, both for economic and ecological reasons. The effects of climate change and the importance of controlling greenhouse gas emissions was covered in Chapter 1 but will be covered here briefly to indicate where climate change ties in with sustainability and sustainable development.

If we continue with our current consumption of resources and usage of fossil fuels, greenhouse gas (GHG) emissions are likely to treble by the year 2100 (Stern, 2007). It is predicted that this increase would cause a rise in global temperature of 5 degrees Celsius, with a 20% chance of more significant temperature increases. The impact of this rise cannot be ascertained, as this degree of warming is outside the experience of human civilisation. Current analysis models can only predict the effects of 2 to 3 degrees warming.

As described in Chapter 1, building construction accounted for 10% of GHG emissions in 2000 (Stern, 2007). By 2018, this had increased to 11% of GHG emissions (refer Figure 1.3, repeated here as Figure 2.3 for clarity), indicating an increase of 1% (IEA, 2019).

(The 28% CO_2 emissions attributed to 'Buildings' in Figure 2.3 refers to the emissions generated by buildings during their lifetime, also referred to as operational carbon emissions)

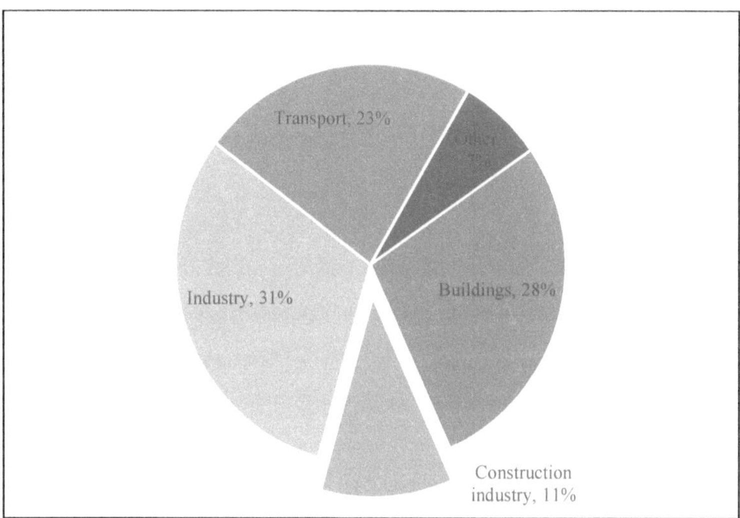

Figure 2.3: Global CO_2 emissions by sector in 2018 (IEA, 2019)

In South Africa, the embodied GHG emissions generated by building construction in 2006 accounted for 5% of the total national annual GHG emissions (Ampofo-Anti, et al., 2015). By contrast, in 2018, industry-related emissions accounted for 16% of the national GHG emissions (Climate Transparency, 2019), with steel production being a significant contributor.

Reducing the carbon footprint of buildings is one way in which to achieve the aim of the Paris Agreement of reducing GHG emissions, and thus mitigating the impacts of climate change.

The impacts of climate change will vary from region to region. The predictions for South Africa are higher temperatures and evaporation rates, both increased and decreased levels of precipitation and severe weather events such as storms (UNU-WIDER, 2016).

2.2.3 Principles of Sustainable Development and Sustainable Construction

The United Nations has developed 17 Sustainable Development goals. The World Green Building Council considers that green building (or sustainable construction) can assist in meeting nine of these goals (refer Figure 2.4).

Figure 2.4: Sustainable Development Goals for green buildings (WGBC, 2016)

Kibert (1994) proposed six principles of sustainable construction and adding a seventh principle in his later works ((Kibert, 2016). Table 2.2 indicates how these principles match the sustainable development goals of the United Nations.

Table 2.2: Principles of Sustainable Construction (adapted from Kibert, 2016)

Principles of Sustainable Construction	Sustainable Development Goal
1. Reduce resource consumption (reduce)	Goal 12: Responsible consumption and production
2. Reuse resources (reuse)	Goal 12: Responsible consumption and production
3. Use recyclable resources (recycle)	Goal 13: Climate action
4. Protect nature	Goal 15: Life on Land
5. Eliminate toxics	Goal 13: Climate action
6. Apply life-cycle costing (economics)	Goal 8: Decent work and economic growth
7. Focus on quality	Goal 9: Industry, innovation and infrastructure

- *Resource Efficiency*

The first three principles as proposed by Kibert (reduce, reuse, recycle) can be considered together as 'resource efficiency'.

In simplistic terms, resource efficiency can be described as "doing more with less for longer" (Smith, et al., 2007). More efficient activities are those that use fewer inputs such as energy, water or materials, while providing the same or better product or service.

Sustainability should be considered from the start of the design phase, at the time when material selection occurs. The aim is to reduce the environmental impact of building construction by minimising the resources used in material production. Suitable and applicable materials can be selected based on the specific criteria and site concerned, with the first choice being locally available, indigenous materials.

Another aspect of resource efficiency is the reuse or recycling of materials. Construction activities use approximately 40% of all raw materials extracted from the Earth each year (NAHB, 2006). At the same time, with remodelling and deconstruction of old buildings, it is estimated that the European Union generated 333 million tonnes of construction waste in 2014 (Menegaki & Damigos, 2018). 8.3% of all waste generated in South Africa in 2017 was classified as deriving from construction and demolition activities, which equates to approximately 4,5 million tons (Department of Environmental Affairs, 2018).

- *Protection of Nature*

Renewable construction materials are also called natural materials. These are often lower in embodied energy (refer Section 2.5.4) than man-made materials, requiring less processing. Some examples of renewable or natural building materials are earth, straw and bamboo.

By using building material that is sourced locally, the environmental impact of transport of the materials is minimised. In addition, when local materials are used in their natural form, they are better suited for local climatic conditions.

The raw material requirements of industrialised countries impact ecosystems, in extraction where landscapes can deteriorate and erosion can set in, in processing and transportation with the emission of greenhouse gases, and in waste disposal where toxic substances could be released into the environment. The type of raw material affects the extent of deterioration. For instance, heavy metals could have a greater environmental impact during their use and end-of-life phase as a result of their poor biological degradability. In a similar vein, if a renewable material such as timber or bamboo is harvested unsustainably, the local ecosystems will be damaged (van der Lugt, et al., 2009)

- *Economics*

Basing the selection of materials on their environmental impact is not always easy, due to the conflicting information available. The most common process or method used is Life Cycle Analysis (LCA) (refer Section 2.5.2).

Building designs and material choices should aim to minimise degradation caused by weathering, and thus reduce maintenance on the structure. If materials are chosen for their durability, and the building design adapted to protect the materials, this will enhance their useful life and minimise maintenance required.

- *Elimination of toxics*

Some materials, such as heavy metals, can be harmful to human health (van der Lugt, et al., 2009). In addition, substances used in the processing of biotic materials can be harmful, such as preservatives used for timber, e.g. arsenic, copper, chrome.

2.2.4 Benefits of Sustainable Construction

The benefits of sustainable construction can be categorised into three areas that correspond to the interlocking ring model for sustainable development, as discussed in Section 2.2.1, namely environmental, economic and social benefits (Stubbs, 2008), (SALGA, 2017), (Oguntona, et al., 2019).

1) Environmental
   - Sustainable construction uses processes and products with a lower environmental impact.
   - The whole life span of a building/project is considered, from construction through operation to the demolition/end-of-life phase.
2) Economic
   - Sustainable construction may bring cost savings, although this would be dependent on what methods and materials were used.
   - Attention to the building design, construction and operation can reduce the overall cost of a building throughout its lifetime through the reduced use of water and energy in the operational phase.
   - Sustainable buildings, structures and infrastructure provide better living and working environments, which in turn help to improve productivity, which in turn encourages economic prosperity.
   - Green buildings are considered to have a higher market value with associated higher rental rates.
3) Social
   - Sustainable construction creates better environments within which people can work and live, resulting in improved levels of well-being and health.

2.3   Current building codes, guidelines and legislation - Globally and in South Africa

A building code is "a set of rules that specify the minimum standards for constructed objects such as buildings" (Anon., 2016). In the past, building codes focussed on safety and durability, whereas nowadays the building codes address energy and environmental aspects as well.

In some countries, building codes are developed and enforced by government agencies or standards organisations; these are known as national building codes. In other countries, the local authorities produce the building codes.

In Europe, the Eurocode has superseded the older national building codes and is now used in most European countries, with each country having a National Annex to localise the contents of the Eurocode.

In 1994, the United States of America formed the International Code Council (ICC) to develop and maintain building codes applicable to commercial, institutional and residential structures. The ICC has 15 codes, covering various aspects. For instance, the International Building Code (IBC) addresses health and safety for buildings based on prescriptive and performance-related requirements. The ICC also has the International Green Construction Code (IgCC) which provides guidelines for sustainable buildings.

In 2013 in the United Kingdom, the government issued Building Regulations Part L, which refers to the conservation of fuel and power, to minimise carbon emissions of buildings.

The National Building Regulations and Building Standards Act, Act 103 of 1977, provides the basis for how buildings in South Africa should be constructed and become enforceable as law in 1985. Shortly thereafter, in 1987 the South African Bureau of Standards (SABS) published the Code of Practice for the Application of the National Building Regulations, SABS 0400-1987. This has since been revised and is now known as SANS 10400. It sets out prescriptive provisions that are deemed to satisfy the technical aspects of the National Building Regulations.. The South African Bureau of Standards (SANS) is responsible for developing standards for the building industry in line with the regulations.

Part X of SANS 10400 deals with environmental sustainability, while Part XA deals with energy usage in buildings. Only Part XA has been published to date.

In South Africa, the local municipalities administer building approval. Since the National Building Regulations (NBRs) came into force, the local authorities have used these as a basis for approval when checking building plans. Conventional construction using concrete, clay bricks, or structural steel framing, is covered by the deemed-to-satisfy rules accompanying the building regulations or rational design. As the NBRs are performance-orientated, a local authority may approve a different construction material based on their knowledge and experience, or the performance of similar buildings in similar conditions.

When a building system or material falls outside the experience of the local authority, they will require assurance of fitness for purpose of the method or material. This assurance can be provided by an applicable report by the Council for Scientific and Industrial Research (CSIR) or South African Bureau of Standards (SABS), the submission of a current certificate by Agrément South Africa, or a rational design by a professional engineer or an approved competent person (Agrément South Africa, 2010).

South Africa has several municipal policies and guidelines that encourage green building practice, a few of which are listed in Table 2.3.

Although the building codes address such issues as energy and resource efficiency, following sustainable construction guidelines is not mandatory. Thus, architects and engineers sit with the issue of convincing their clients of the benefits of such construction, despite the initial higher capital outlay.

Table 2.3: South African municipal policies and guidelines for Green Building (SALGA, 2017)

Municipality	Policy/Guideline
Cape Town, Western Cape	Smart Building Handbook, 2012
Drakenstein, Western Cape	Green Building Manual, 2010
Johannesburg, Gauteng	Design Guidelines for Energy Efficient Buildings, 2008
KwaDakuza, KwaZulu-Natal	Green Building Guidelines, 2015
Msunduzi, KwaZulu-Natal	Green Building Guidelines, 2015
Steve Tshwete, Mpumalanga	Green Building Guidelines, 2015
Tshwane, Gauteng	Green Building Policy and By-law, 2009

2.4 Costs of Sustainable Construction and Green Building

There is a widespread perception that green buildings cost more than conventional buildings, both in the design and the construction thereof (Kats, 2003). In 2014, a tender was put out for the construction of a school in Bonnievale, in the Western Cape of South Africa. This school included green building elements such as solar-powered panels and a thermal rock store below the school hall to omit the need for artificial ventilation. The tender prices averaged at R 75 million, which was approximately 52% higher than the Quantity Surveyor's estimate. When those elements were removed, the eventual building cost was R 55 million (P du Plessis 2014, personal correspondence, 5 May).

The additional cost was attributed to the contractors' lack of experience with such sustainable technologies which resulted in higher premiums for the work. Furthermore, increased control and supervision costs were necessary to meet the Green Star requirements, which were stated as 30% higher than conventional building costs. Lastly, the location of the project added to the cost as this necessitated additional transport costs for certain specified materials.

However, green building has long-term operational benefits such as energy and water savings, increased health of occupants with a resultant increase in productivity, and reduced maintenance costs.

The Green Building Council South Africa (GBCSA), together with the Association of South African Quantity Surveyors and the University of Pretoria compiled a report documenting the costs and trends of green buildings constructed in South Africa. In their initial report published in 2016, they indicated that the average green cost premium of office projects was 5,2%, where the green cost premium is defined as "the additional cost of green building over and above the cost of conventional construction, expressed as a percentage of the total cost of the project". The updated report published in 2019 indicated that this premium had dropped to 3,9% (GBCSA, 2019). It also indicated that Green Star certified offices reported a 1% lower electricity usage and a 24% lower water usage per occupied square metre.

Thus, this research indicates that the initial construction costs of green or sustainable construction are currently higher than conventional construction. However, the long-term savings of such building offset the initial higher capital outlay.

2.5 Tools for Measuring Sustainability

There are several tools available for measuring or rating the sustainability of buildings. The choice of tool or method is dependent on the end-goal of the assessment. If the goal is to obtain a certificate or green rating, perhaps for commercial reasons, then one of the rating systems would seem appropriate. If the focus is on material selection, then a Life Cycle Analysis or Carbon footprint would be more appropriate.

Another deciding factor could be the time and effort required for the various tools. The rating systems use points or credits, and are simple to complete, requiring minimal time and effort. On the other hand, a Life Cycle Analysis requires an in-depth research of the materials, from extraction through to production, and possibly to end of life or demolition of the building. This research would take more time and effort than a rating system.

2.5.1 Rating Systems

Environmental tools and rating systems have been established in various countries, which assess environmental performance against a set of quantitative and qualitative criteria. Most rating systems work on a credit or point system, where credits or points are assigned.

The first rating system was developed by the Building Research Establishment in the United Kingdom, known as the Building Research Establishment Environmental Assessment Method (BREEAM), and presented a systematic method to evaluate the implementation and performance of green buildings (Zhang, et al., 2019). Subsequently, America established the Leadership in Energy and Environmental Design (LEED) certification system, which was later adopted by Brazil and adapted to the Brazilian market. Japan uses the Comprehensive Assessment System for Building Environmental Efficiency (CASBEE), and France uses the High Environmental Quality (HEQ) system. Australia established the Green Star rating system in 2003, which has since been adopted by New Zealand, while the Green Star SA rating system in South Africa was based on the Australian Green Star. South Africa also uses the Sustainable Building Assessment Tool (SBAT) (World Green Building Council, 2009), (Wang & Adeli, 2014), (Muigai, 2014) (refer Table 2.4).

These rating systems provide a scoring system, whereby points or credits are obtained in different areas, and the total number of points achieved will determine how "green" the building is considered to be. Each rating system has its own credits, weightings and requirements to achieve certification, making it difficult to compare the systems.

Table 2.4: Current global building rating systems and tools

Country	Association	Rating System
United States of America	U.S. Green Building Council	LEED (Leadership in Energy and Environmental Design)
United Kingdom	Building Research Establishment	BREEAM (Building Research Establishment Environmental Assessment Method)
Japan	Japan Sustainable Building Council	CASBEE (Comprehensive Assessment System for Building Environmental Efficiency)
France	Association HQE	HQE (Haute Qualité Environnementale)
Germany	German Sustainable Building Council	DGNB (Deutsche Gesellschaft für Nachaltiges Bauen)
Australia, New Zealand	Green Building Council of Australia	Green Star
South Africa	CSIR South Africa	Green Star SA
	South Africa Green Building Council	SBAT (Sustainable Building Assessment Tool)

One of the rating systems used in South Africa is described below as an example.

- *Sustainable Building Assessment Tool (SBAT):*

The Sustainable Building Assessment Tool (SBAT) rating system was developed by the Council for Scientific and Industrial Research (CSIR) to measure the performance of buildings in terms of their contribution to sustainable development, with particular focus on a developing country context, as well as social and economic criteria. It has been developed to be suitable for a wide range of building types, such as schools, light industrial buildings, service stations, residential housing and offices.

Existing buildings and designs are evaluated according to criteria given in the tool. Points are awarded and performance is reflected as a figure out of 5 for each area, as well as an overall sustainability rating. The tool can also indicate target values for each criteria, which can be used to assess improvement of the building or design in those particular areas.

An example of an SBAT report is shown in Figure 2.5. The report provides a visual representation of areas where improvement in environmental performance could be achieved (Gibberd, 2008), (Gibberd, 2017)

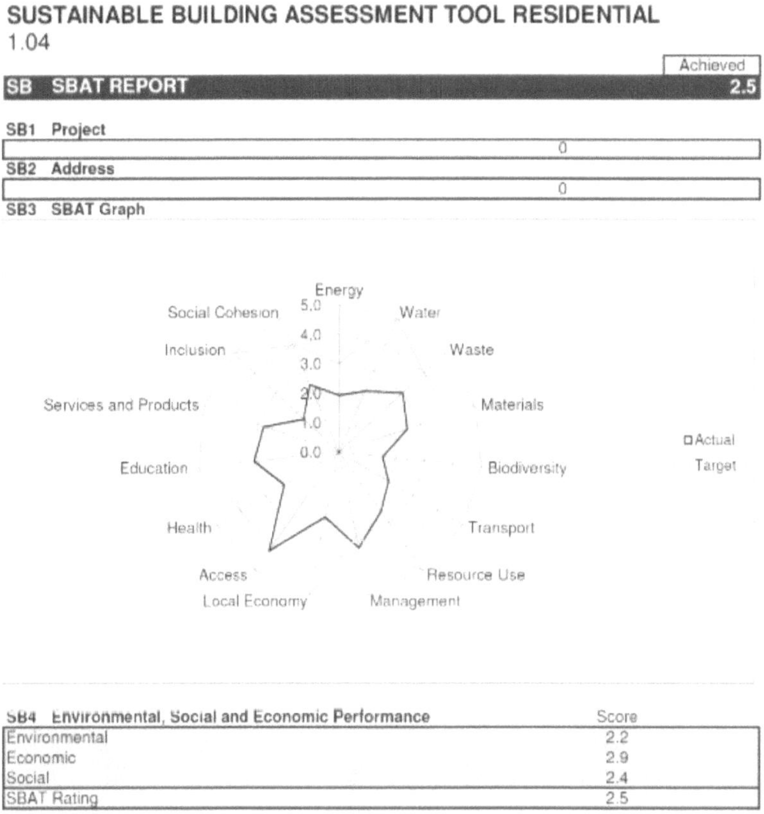

SUSTAINABLE BUILDING ASSESSMENT TOOL RESIDENTIAL 1.04

		Achieved
SB	**SBAT REPORT**	**2.5**

SB1 Project

0

SB2 Address

0

SB3 SBAT Graph

SB4 Environmental, Social and Economic Performance	Score
Environmental	2.2
Economic	2.9
Social	2.4
SBAT Rating	2.5

Figure 2.5: Example of an SBAT rating system report (Gibberd, 2017)

2.5.2   Life Cycle Analysis

Life Cycle Analysis (LCA), also known as Life Cycle Assessment, is a commonly accepted basis for environmental assessment of products (van der Lugt, et al., 2003), (van den Dobbelsteen, 2004). In such an assessment, all the environmental effects that occur during the life cycle of a product, from extraction to the end phase of its life cycle, either demolition or recycling, are systematically assessed and analysed. This assessment is sometimes referred to as 'cradle to grave' assessment. LCA is based on several environmental indicators such as carbon footprint, acidification, eutrophication, toxicity, depletion, land-use and waste. It is used to indicate to what extent a product or material has an impact on the environment.

The LCA methodology was developed by the Centre of Environmental Studies in Leiden, Netherlands, in 1992. The standards for methods and procedures were developed and standardised by the International Standards Organisation to form the ISO 14040 series.

The standard LCA uses a functional unit to compare alternatives on a common or general basis. Measurement of the alternative products being compared are defined by their technical and functional requirements during a certain lifespan. Maintenance and repair are included in the functional unit, so that weaker alternatives which require more material, and alternatives with a shorter lifespan, will lead to a heavier environmental load.

LCA includes environmental effects that can be quantified, excluding effects such as the deterioration of ecosystems. It does not include a weighting method, which makes it difficult to compare different products based solely on the LCA results. Additional methods are used for these, such as the TWIN model, developed by Michael Haas, Director of the Dutch Institute for Building Biology and Ecology, in 1997 (van der Lugt, et al., 2005) or the Environmental costs (Eco-costs) methods (van den Dobbelsteen, 2004). The TWIN model uses LCA as its basis but includes for qualitative evaluations of environmental aspects that are not quantifiable, such as health and deterioration aspects. It includes a weighting factor for each aspect assessed, based on the principle of environmental costs or eco-costs (refer Section 2.5.3), and as such reduces the 14 criteria from the LCA analysis into one total score of one figure for comparison.

TWIN[2002], the successor to the TWIN model developed by Haas, does not weight the various effects but adds a multiplication factor based on environmental cost per effect, which can be summed to obtain and end-performance value (van der Lugt, et al., 2003).

The most significant potential for emissions reductions on a building occurs during the early design phase, when the project data for conducting a detailed LCA may not be available. However, there are simplified tools available, specifically developed for these early design-stage assessments. For projects that use building information modelling (BIM), BIM-integrated tools provide an ongoing assessment of the effect of design changes on the embodied carbon results. The availability and use of such tools vary by project, region and country. Care should be taken when using tools from other regions or countries, as the underlying LCA datasets associated with a particular tool may only be representative of that country or region, due to the differences in the way that materials and products are sourced and made in the different regions.

Examples of such tools include the French ELODIE tool; eToolLCD and The Footprint Calculator from Australia; One Click LCA in Finland and Tally, a USA developed plug-in for the design software Revit (World Green Building Council, 2019).

2.5.3   Environmental costs (Eco-costs)

In the building and construction industry, LCA's are characterised by many indirect environmental impacts, which cannot be measured. The environmental or eco-cost/value model is an attempt to assign costs to each ecological impact (environmental costs), which can then be used as a cost comparison between various products (Vogtlander, 2001).

Eco-costs are 'virtual' costs, i.e. costs relating to measures that must be taken to recycle or reuse a product in order to not exceed the Earth's estimated carrying capacity. Eco-costs have been defined by Hendricks (2006) as follows:

- Virtual pollution prevention costs, i.e. the costs involved to reduce the emissions in the product chain to a sustainable level
- Energy eco-costs, i.e. the costs for sustainable energy sources
- Material depletion eco-costs, i.e. the costs of the proportion of raw materials that cannot be recycled (Hendriks, et al., 2006)

2.5.4   Embodied Energy

Embodied energy refers to the amount of energy required by all processes associated with the design, manufacture and supply of a product or material to the point of use (Stubbs, 2008). For buildings, this would cover the design and construction phase as well as the manufacture of all individual building components and the energy used by transportation and machinery.

There are several tools to assess the embodied energy of materials. The most common one is the Inventory of Carbon and Energy database, produced by the University of Bath. This database was created and first released in 2005, providing embodied energy and embodied carbon factors for common building materials, such as concrete, timber and steel. Since 2019, the embodied energy factors are no longer included as the data in the literature, which the ICE database relies upon, generally uses embodied carbon as the primary reporting metric (refer Figure 2.6). The perception is that carbon emissions provide a better indicator of the effect on global warming and climate change. The ICE database defines embodied carbon as the sum of fuel and process-related carbon emissions from extraction to the point of use, also known as cradle-to-gate (ICE, 2019). This is also referred to as the carbon footprint of a product or material (refer Section 2.5.5).

Figure 2.6: Embodied Energy and Embodied Carbon values for various materials (ICE, 2019)

2.5.5 Carbon Footprint

There are several definitions for the term 'carbon footprint', most using this term as a generic synonym for carbon dioxide or greenhouse gas emissions expressed in CO_2 equivalents (CO_2-eq). Wiedmann (2008) performed a literature review of scientific journals covered by Scopus and ScienceDirect, two online databases of peer-reviewed research literature, for the term 'carbon footprint'. A selection of his results is presented in Table 2.5.

Table 2.5: Definitions of 'carbon footprint' from literature (Wiedmann & Minx, 2008)

Definition	Source
"…a methodology to estimate the total emission of greenhouse gases (GHG) in carbon equivalents from a product across its life cycle from the production of raw material used in its manufacture, to disposal of the finished products (excluding in-use emissions)." "…a technique for identifying and measuring the individual greenhouse gas emissions from each activity within a supply chain process step and the framework for attributing these to each output product."	Carbon Trust (2007)
"…the 'Carbon Footprint' is a measure of the impact human activities have on the environment in terms of the amount of greenhouse gases produced, measured in tonnes of carbon dioxide."	ETAP (2007)
"The demand on biocapacity required to sequester (through photosynthesis) the carbon dioxide (CO_2) emissions from fossil fuel combustion."	Global Footprint Network (2007)
"A 'carbon footprint' is the total amount of CO_2 and other greenhouse gases, emitted over the full life cycle of a process or product. It is expressed as grams of CO_2 equivalent per kilowatt hour of generation (gCO2-eq/kWh), which accounts for the different global warming effects of other greenhouse gases."	Parliamentary Office of Science and Technology (POST 2006)

Wiedmann further provided his own definition of carbon footprint, namely "a measure of the exclusive total amount of carbon dioxide emissions that is directly and indirectly caused by an activity or is accumulated over the life stages of a product."

Due to the variability in definitions, where the carbon footprint of a product or building material is given, it would be necessary to clarify what was included or excluded from that assessment, particularly if this is used as a basis for comparison between different products or materials.

2.6 Bamboo as a sustainable construction technology

One of the aspects of sustainable construction where the engineer can play a significant role is in the selection of appropriate materials and construction technologies. In many countries, such as India, China and Brazil, bamboo construction is a common and accepted form of construction. However, the scientific and engineering literature on this type of construction is minimal, with a large proportion of the literature being provided by architects.

Bamboo has been touted as a preferable building material in recent times in terms of sustainability and ecological impact. It is reported to strong and flexible, but lightweight. Bamboo plants meet the criteria for sustainable, renewable construction materials in terms of growth rate and processing, and many suitable varieties are available for construction (van der Lugt, et al., 2009). They reach maturity in 3 to 6 years, from when they become suitable for use as structural members in construction.

2.7 Conclusions

The concept of sustainability in engineering was first documented in 1925 by Professor Arrhenius. Since then, the concept developed until it became part of the environmental terminology in 1987 with the Brundtland Report. This report is seen to be the starting point for sustainable development and produced the traditionally accepted definition of sustainable development, which is "development that meets the needs of the present without compromising the ability of future generations to meet their own needs" (Mebratu, 1998).

Since the publication of the report, numerous conferences have been held on sustainability issues, with the publication of further reports, such as the Kyoto Protocol and Agenda 21. One of the results of these conferences was a 10-year framework of programmes on sustainability, including protecting and restoring the Earth's ecosystems, reducing the emission of pollutants and promoting the efficient use of natural resources and products.

In this chapter, various definitions were reviewed for the concepts of sustainability, sustainable development and sustainable construction. Although the definitions vary in detail, the general theme for the definition of sustainable construction concerns the protection of the natural environment. The simplest definition was offered by Singh (2007), who defined sustainable construction as a suitable choice of materials, sources, construction methodologies and design philosophies to increase performance, reduce environmental impact and waste, and to be less harmful to the environment.

One area where construction is affecting the environment is the release of greenhouse gases (GHG) to the atmosphere. In 2018, building construction was responsible for 11% of GHG emissions worldwide. Reducing the carbon footprint of buildings would be one way to reduce the GHG emissions, and thus to mitigate the effects of climate change.

There are several tools available for measuring the sustainability of buildings. One of the most common tools for the assessment of materials is the Life Cycle Analysis (LCA), also known as Life Cycle Assessment. However, in the construction industry, LCA's are characterised by many indirect environmental impacts, which cannot be measured by such a system. Measurement of embodied energy and the carbon footprint are two ways whereby these indirect impacts can be quantified so that alternative materials can be compared in terms of their level of sustainability.

Bamboo is considered to be a building material with a low environmental impact. The following chapters
in this book will explore the various aspects of the bamboo plant to determine whether it is indeed suitable for sustainable construction.

2.8   References

Agrément South Africa, 2010. *Technical assessment of construction products in South Africa.* [Online] Available at: http://www.agrement.co.za/pgcontent.php?UID=482 [Accessed 17 March 2010].

Ampofo-Anti, N., Dumani, N. & Van Wyk, L., 2015. *Potential for reducing greenhouse gas emissions in the South African Construction sector,* Pretoria: CSIR.

Anon., 2016. *Building Code.* [Online] Available at: http://en.wikipedia.org/wiki/ Building_code [Accessed 20 January 2016].

Brundtland, G., ed., 1987. *Our Common Future: The World Commission on Environment and Development.* Oxford: Oxford University Press.

Climate Transparency, 2019. *Brown to Green: The G20 transition towards a net-zero emissions economy,* Berlin: Climate Transparency.

Department of Environmental Affairs, 2018. *South Africa State of Waste - A report on the state of the environment,* Pretoria: Department of Environmental Affairs.

GBCSA, 2019. *Green Building in South Africa - Guide to Costs and Trends,* Pretoria: Green Building Council of South Africa, Association of South African Quantity Surveyors, University of Pretoria.

Gibberd, J., 2008. *The sustainable building assessment tool: integrating sustainability into current design and building processes,* Melbourne, Australia: World Sustainable Building Conference.

Gibberd, J., 2017. A Critical Evaluation of the Sustainable Building Assessment Tool. *West Africa Built Environment Research Conference,* 16-17 August 2017.

Hayles, C. S. & Holdsworth, S. E., 2008. Curriculum Change for Sustainability. *Journal for Education in the Built Environment,* July, Vol 3(1), pp. 25-48.

Hendriks, C. F., Vogtlander, J. G. & Janssen, G. M., 2006. The Eco-costs/Value Ratio: A Tool to determine the Long-term strategy for delinking Economy and Environmental Ecology. *International Journal of Ecodynamics,* 1(2), pp. 136-148.

ICE, 2019. *Embodied Carbon - The ICE Database.* [Online] Available at: http://www.circularecology.com/ice-database.html

IEA, 2019. *2019 Global Status Report for Buildings and Construction: Towards a zero-emission, efficient and resilient buildings and cosntruction sector,* s.l.: Global Alliance for Buildings and Construction.

IPCC, 2018. *Global Warming of 1.5°C,* s.l.: IPCC.

Jabareen, Y., 2008. A New Conceptual Framework for Sustainable Development. *Environment Development and Sustainability,* Volume 10, pp. 179-192.

Kates, R. W., Paris, T. M. & Lieserowitz, A. A., 2005. What is Sustainable Development?. *Environment: Science and Policy for Sustainable Development,* April, 47(3), pp. 8-21.

Kats, G. H., 2003. *Green Building Costs and Financial Benefits,* Massachusetts, USA: Massachusetts Technology Collaborative.

Kibert, C., 2016. *Sustainable Construction - Green Building Design and Delivery.* 4th ed. Hoboken, New Jersey: John Wiley & Sons, Inc.

Kibert, C. J., 1994. *Establishing principles and model for sustainable construction.* Tampa, Florida, University of Florida, pp. 200-210.

Kibert, C. J., Sendzimir, J. & Guy, G. B., 2002. *Construction Ecology - Nature as the basis for green buildings.* London: Spoon Press.

Landman, M., 1999. *Breaking through the Barriers to Sustainable Building,* s.l.: s.n.

Leggett, J. & Carter, N., 2012. *Rio+20: The United Nations Conference on Sustainable Development, June 2012,* Washington,D.C: Congressional Research Service.

Lélé, S. M., 1991. Sustainable Development: A Critical Review. *World Development,* 19(6), pp. 607-621.

Mebratu, D., 1998. Sustainability and Sustainable Development: Historical and Conceptual Review. *Environmental Impact Assessment Review,* Volume 18, pp. 493-520.

Menegaki, M. & Damigos, D., 2018. A review on current situation and challenges of construction and demolition waste management. *Current Opinion in Green and Sustainable Chemistry,* Volume 13, pp. 8-15.

Muigai, R. N., 2014. *A framework towards the design of more sustainable concrete structures,* Cape Town: University of Cape Town.

NAHB, 2006. *NAHB Model Green Home Building Guidelines,* Washington, D.C.: National Association of Home Builders.

Oguntona, O. et al., 2019. Benefits and Drivers of Implementing Green Building Projects in South Africa. *Journal of Physics: Conference Series,* Volume 1378.

RICS, 2010. *Global Glossary of Sustainability Terms.* [Online] Available at: http://www.trentglobal.edu.sg/wp-content/uploads/2016/09/RICS-Sustainability-Glossary-June-2010.pdf [Accessed 2015].

SALGA, 2017. *Leadership in Green Building.* [Online] Available at: https://www.salga.org.za/Documents/Knowledge%20Hub/Publications/Green%20Building%20Bookl et%202017.pdf [Accessed 15 July 2020].

Singh, T. P., 2007. Sustainable Construction. *Sustainability Tomorrow,* July.

Smith, M., Hargroves, K., Desha, C. & Palousis, N., 2007. *Engineering Sustainable Solutions Program: Critical Literacies Portfolio - Introduction to Sustainable Development for Engineering and Built Environment Professionals,* Australia: The Natural Edge Project (TNEP).

Stern, N., 2007. *Stern Review: The Economics of Climate Change,* Cambridge: Cambridge University Press.

Stubbs, B., 2008. *Plain English Guide to Sustainable Construction,* London: Constructing Excellence.

The Institute of Structural Engineers, 1999. *Building for a sustainable future: Construction without depletion.* s.l.:SETO.

UNU-WIDER, 2016. *Future climate scenarios for South Africa: The importance of mitigating emissions. WIDER Research Brief 2016/6.* Helsinki: UNU-WIDER.

Valdes, H., Correa, C. & Mellado, F., 2018. Proposed Model of Sustainable Construction Skills for Engineers in Chile. *Sustainability,* 30 August.Volume 10.

van den Dobbelsteen, A., 2004. *The Sustainable Office - an exploration of the potential for factor 20 environmental improvement of office accommodation,* Delft: Technische Universiteit Delft.

van der Lugt, P., van den Dobbelsteen, A. & Abrahams, R., 2003. Bamboo as a building material alternative for Western Europe? A study of the environmental performance, costs and bottlenecks of the use of bamboo (products) in Western Europe. *Journal of Bamboo and Rattan,* 2(3), pp. 205-223.

van der Lugt, P., van den Dobbelsteen, A. & Janssen, J., 2005. *An environmental, economic and practical assessment of bamboo as a building material for supporting structures.* [Online] Available at: https://www.sciencedirect.com/science/article/abs/pii/S0950061805001157 [Accessed 03 June 2011].

van der Lugt, P., Vogtländer, J. & Brezet, H., 2009. *Bamboo, a Sustainable Solution for Werstern Europe: Design Cases, LCAs and Land-use,* Beijing: INBAR.

Vogtlander, J., 2001. *The model of Eco-costs/Value Ratio - A new LCA based decision support tool,* Gronigen, Netherlands: Delft University of Technology.

Wang, N. & Adeli, H., 2014. Sustainable Building Design. *Journal of Civil Engineering and Management,* November, 20(1), pp. 1-10.

WGBC, 2016. *Green building Sustainable Development Goals [image].* [Online] Available at: https://www.worldgbc.org/green-building-sustainable-development-goals [Accessed 25 August 2020].

Wiedmann, T. & Minx, J., 2008. A Definition of 'Carbon Footprint'. In: C. Pertsova, ed. *Ecological Economics Research Trends.* Hauppauge NY: Nova Science Publishers, pp. 1-11.

World Green Building Council, 2009. *Six Continents, One Mission,* s.l.: World Green Building Council.

World Green Building Council, 2019. *Bringing embodied carbon upfront,* London: World Green Building Council.

Zhang, Y. et al., 2019. A Survey of the Status and Challenges of Green Building Development in Various Countries. *Sustainability,* 11(19).

Chapter 3

3 BAMBOO AS A PLANT

3.1 Introduction

Recognition of bamboo as a sustainable construction material has increased over the past two decades. Following the principles of sustainable construction as discussed in the previous chapter, it can be seen to fulfil many of these principles, such as resource efficiency and the use of renewable materials. These will be discussed in more detail in Chapter 4.

There are many varied structural applications of bamboo, which include flooring boards, particleboards, laminated beams, reinforcement for concrete, and framing. However, most knowledge of bamboo design and construction has been based on cultural tradition and bamboo products, and to date has largely been addressed by the architectural profession. In order to develop bamboo as a sustainable construction material, and to promote its use in the engineering profession, the traditional building techniques should be addressed in terms of engineering standards, as well as the evaluation and understanding of traditional building techniques and the context or environment in which they are used. Furthermore, a greater understanding of the engineering properties and performance of bamboo should be developed.

This chapter will address the main botanical aspects, as well as the anatomical structure and properties of the bamboo plant that affect engineering design. The focus will be on the main bamboo species used for construction worldwide and will highlight areas that require further research.

3.2 Botanical Taxonomy and Classification

Plants are classified according to their taxonomical characteristics. The most common classification system is the Cronquist System, which was developed by Arthur Cronquist in 1968, and revised in 1988. According to Cronquist (1988), bamboo is botanically classified as shown in Figure 3.1:

Bamboo is the collective, common name for different species of giant grasses, belonging to the family *Poaceae* (formerly known as *Gramineae*). Each family is divided into subfamilies, each representing a single lineage with distinctive features or characteristics. Thus, for example, the subfamily *Bambusoideae* is characterised by the specific shape of the mesophyll cells in the leaves (Kellogg, 2001), which enable photosynthesis to take place (see Figure 3.2).

The subfamily *Bambusoideae* represents the woody and herbaceous bamboos and is sub-divided into approximately 15 tribes. The largest tribe, *Bambuseae*, refers to the woody bamboos, while the tribe *Olyreae* refers to the herbaceous bamboos, and the tribe *Arundinarieae* refers to the temperate woody bamboos (Kelchner, 2013). Each of the tribes is divided into sub-tribes, which are further divided into multiple genera.

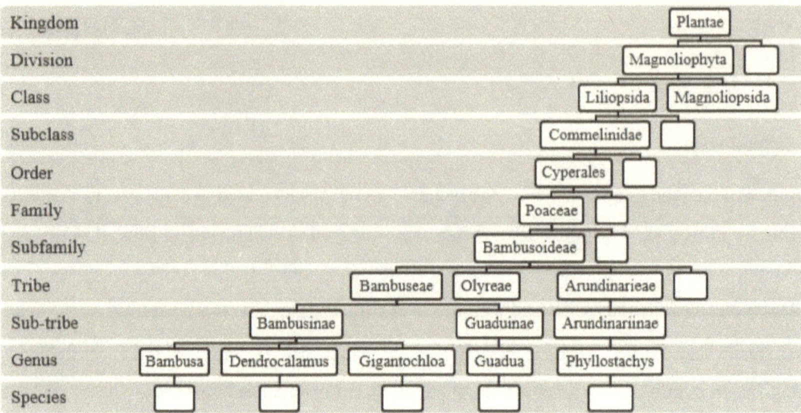

Figure 3.1: Taxonomical classification of Bamboo plants (Cronquist, 1988)

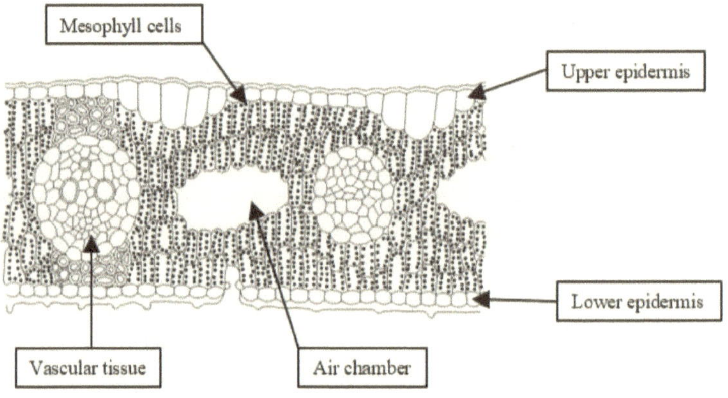

Figure 3.2: Typical cross-section of a bamboo leaf (Kumar, 2017)

3.3   Macrostructure of the Bamboo Plant

Bamboo plants consist of the visible stems and leaves, and the underground root system (refer Figure 3.3). The stem, known as the culm, consists of the branching system, sheath, leaves, flowers, fruit and seedlings. The roots provide the transport of water and nutrients from the soil to the remainder of the bamboo plant (Chaowana, 2013). The shallow root system forms in the top 30 cm of the soil, and in this way forms a mesh that prevents soil erosion.

3.3.1   The Culm (Stem)

The culm or stem is the portion of the bamboo plant found above ground. They are generally straight and hollow and vary in size, diameter and texture, dependent on the species. At particular spacings along the culm, there are transverse diaphragms or nodes, which are solid. These diaphragms provide a pathway for nutrients, as well as preventing buckling and cracking of the walls by providing cross-sectional stiffness. The culm length between the nodes is called the internode (refer Figure 3.4 and Figure 3.5). In the internodes, the cells are oriented axially along the length of the culm, with no radial cell elements. Most internodes are hollow, although some species have solid internodes, often referred to as "male bamboo". Thus, in the species with hollow internodes, the only transverse connection in the culm is at the nodes or diaphragms.

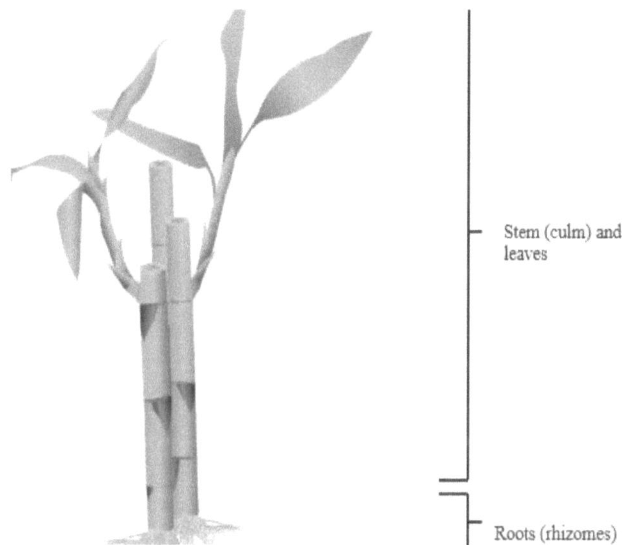

Figure 3.3: Parts of a bamboo plant (Cyberscooty, n.d.)

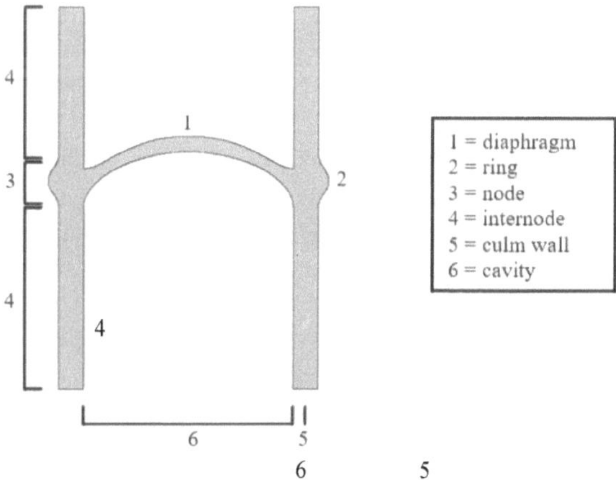

Figure 3.4: Parts of a bamboo culm (Janssen, 2000)

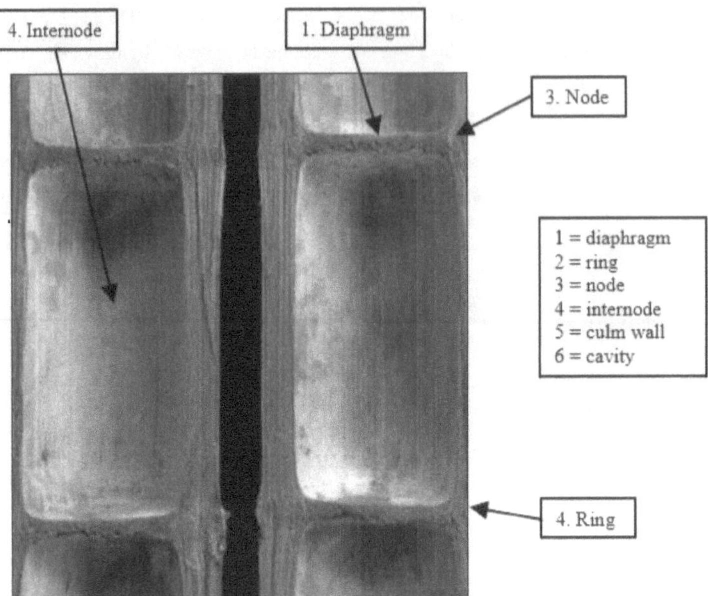

Figure 3.5: Longitudinal section of a culm (Richard, 2013)

The macroscopic characteristics of bamboo vary along the height or length of the culm, with the culm diameter and wall thickness decreasing from the base upwards (Janssen, 2000), as illustrated in Figure 3.6. Although this figure represents research done on four specific species of bamboo, two of which are commonly used for construction purposes, namely *Dendrocalamus asper* and *Phyllostachys pubescens*, it is typical of most species. The cross-section of the culm is roughly circular, tapering towards the top of the plant. Shorter culms taper more strongly, as seen in the culm lengths shorter than 2 m. The longer culm lengths taper less, and as a result, these are often preferred for building purposes, as this results in straighter culms with less change in diameter (Chaowana, 2013). Furthermore, longer length culms require fewer or no joins or laps to create structures and shelters where people can stand upright. However, it should be noted that with the longer culms, although there is less variation in wall thickness, the thickness itself is thinner than with the shorter culms. Thus, it would appear that a compromise is necessary between minimal change in wall thickness and the resulting reduction in strength properties associated with the thinner walls.

In addition to the variations in culm diameter and wall thickness, the number of and length of the internodes is dependent on species. The segmental length of the internodes is largest towards the middle of the culm, and smaller at the base and top.

The outermost skin or layer of the bamboo culm wall is covered with a waxy layer of silica, which protects the culm from water ingress. The young shoots are protected with culm sheaths (refer Figure

3.7), covered with tiny sharp hairs that can injure and irritate human skin. These sheaths are shed by the end of the first year of growth.

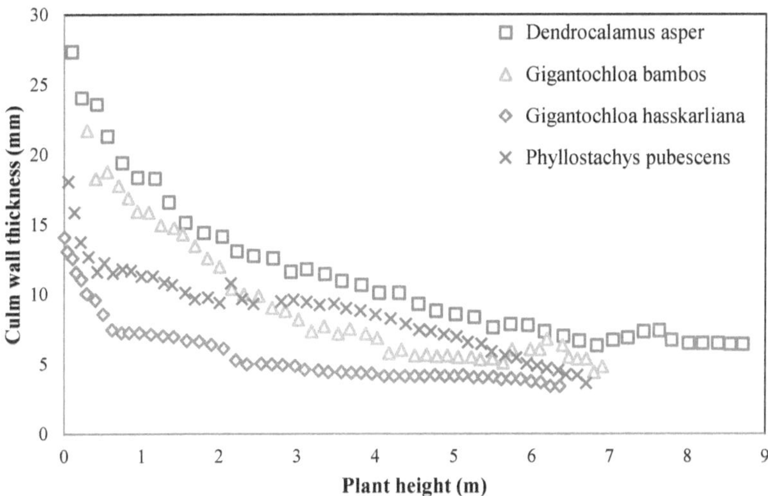

Figure 3.6: Variation of culm wall thickness along the plant height of four bamboo species (Chaowana, 2013)

Figure 3.7: Culm sheaths (Shelton, 2009)

Branches usually form on the culms at the nodes. These branches are also made up of hollow internodes and solid nodes (refer Figure 3.8).

Figure 3.8: Branching of bamboo (Shelton, 2009)

3.3.2 The Rhizomes (Roots)

The rhizome or root structure is the primary structure through which bamboos reproduce. The rhizomes are not actual roots, but instead are underground stems with nodes and internodes that grow laterally (Richard, 2013). Bamboo plants are characterised by their type of root or rhizome formation. Three main groups have been identified, namely monopodial (leptomorph), sympodial (pachymorph) and amphipodial bamboos (Janssen, 2000).

Monopodial, or leptomorph, bamboo plants have long thin rhizomes, with buds that produce shoots at regular intervals from the internodes. New culms or stems grow from these shoots perpendicular to the underground rhizome. These are also known as spreading or running bamboo, with the tip of the rhizome growing horizontal to the ground. There are no shoot buds on the culm base, all shoot buds form on the rhizome itself (refer Figure 3.9 and Figure 3.10).

The running bamboos are considered an invasive species, due to their spreading nature. Some species send underground shoots that extend several meters per year.

Sympodial or pachymorph bamboo plants have short thick rhizomes, with culms that are produced from the tip rather than along the rhizome length. These bamboo culms grow close together as a clump, with the new shoots starting close to the base of the existing culm or stem leading to the term "clumping bamboo" (refer Figure 3.11 and Figure 3.12).

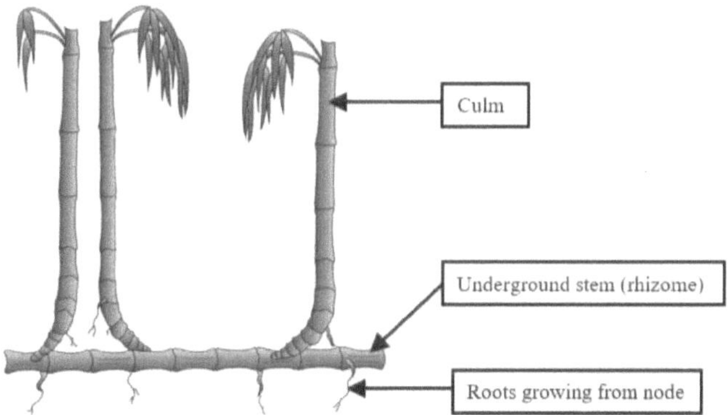

Figure 3.9: Monopodial (leptomorph, running) root diagram (Bamboo Botanicals, 2010)

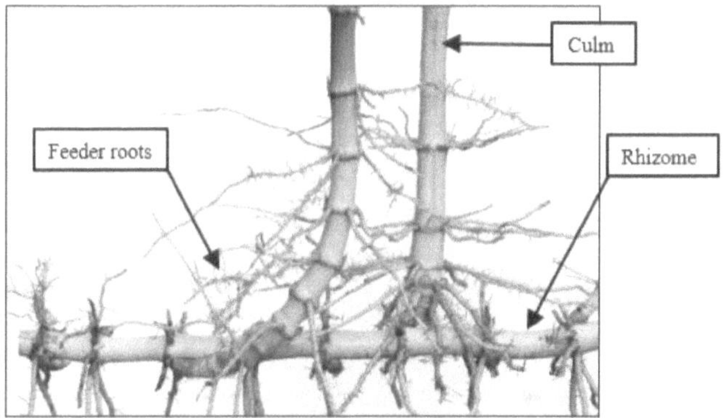

Figure 3.10: Monopodial bamboo (soil removed) (Bamboo Botanicals, 2010)

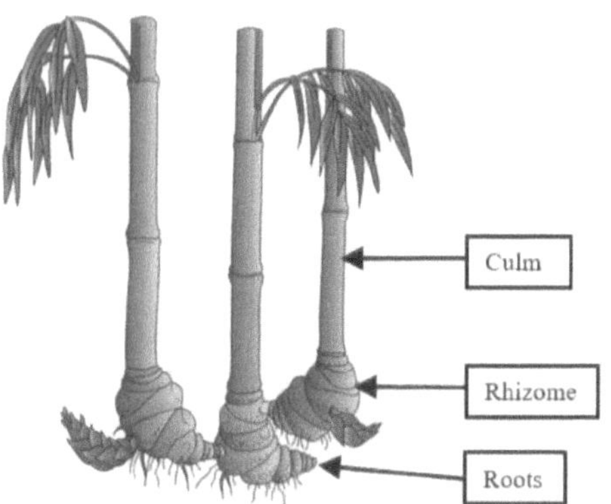

Figure 3.11: Sympodial (pachymorph, clumping) root diagram (Bamboo Botanicals, 2010)

Figure 3.12: Clumping bamboo (Bamboo Botanicals, 2010)

Although clumping bamboo plants are preferred from a growth point of view, as they occupy less land area, their clumping nature can be problematic for harvesting. Some of the culms grow too close together to harvest, and often the innermost culms are inaccessible for harvesting.

A third type of rhizome system, the amphipodial system, shares features of both the sympodial and monopodial rhizome systems (refer Figure 3.13). Culms form either from buds on the culm base (monopodial) or from rhizome buds (sympodial). These form compound groups of culms above ground (Yuming & Chaomao, 2010).

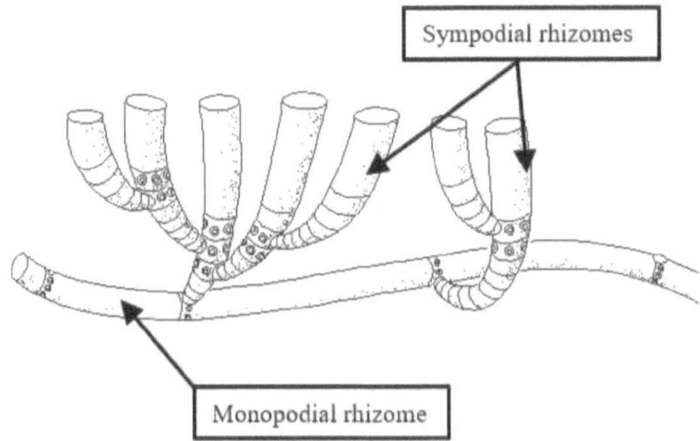

Figure 3.13: Amphipodial compound root formation (Yuming & Chaomao, 2010)

3.4   Microstructure of the Bamboo Plant

The microstructure of a bamboo plant determines its physical and material properties. As the culm or stem is the main part of the plant that is used for construction, only the microstructure of the culm will be reviewed here.

The microstructure of the culm wall varies from the outside to the inside, as shown in Figure 3.14.

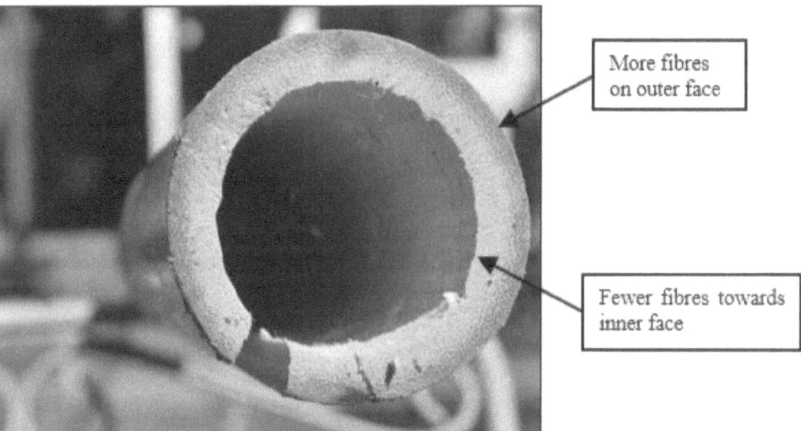

Figure 3.14: Cross-section of a bamboo culm showing variation in fibre density across the culm wall (Shamburg, 2010)

In bamboo, the tissues are formed as follows (Janssen, 1981) and shown in Figure 3.15:

1) *Epidermis* – the inner and outer layers of the culm wall
2) *Parenchyma* – the ground tissue enclosing the vessels and fibres
3) *Sclerenchyma* – the support cells providing strength to the walls
4) The *vascular system* is divided into two parts:
   a. *Phloem* – sieve tubes for the conduction of food materials to the rhizome system.
   b. *Xylem* – the inside of the phloem, made up of vessels, for the transport of water and minerals upwards from the rhizomes
5) *Meristema* - dividing tissue

A typical bamboo culm has approximately 40% fibres (sclerenchyma), 10% vessels (vascular system) and 50% parenchyma (Janssen, 2000). The exact proportions vary according to the position within the culm, age of the bamboo, conditions of growth and species.

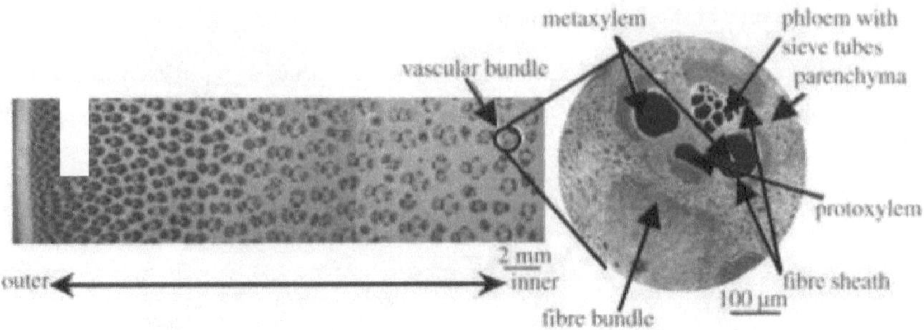

Figure 3.15: Tissues found in a typical bamboo culm (Mannan, et al., 2017)

3.4.1   Epidermis

The outermost portion of the culm wall is known as the *epidermis*. It is approximately 0.25 mm thick, containing a large proportion of silica, making the plant resistant to moisture ingress.

3.4.2   Parenchyma

Parenchyma is the soft cell tissue of plants, as found in the stem pith. The cells are mostly thin-walled and connected by pits. This is the weakest tissue found in bamboo plants.

The function of the parenchyma tissue is to store the food and nutrients in the plant. Both the root system and the culm serve as storage for starch, which acts as an energy source to produce new shoots. The fast growth of shoots depends entirely on the carbohydrates, both starch and sugar as stored in the parenchyma cells. However, starch also makes the bamboo plant susceptible to beetle attack and accelerates fungal deterioration (Liese, 1998). This topic is addressed in more detail in Chapter 4.

3.4.3   Sclerenchyma

Within the parenchyma or matrix are found the cellulose fibres, or *sclerenchyma*, surrounded by lignin. These fibres function as reinforcement or strengthening to the culm wall, and surround the vascular system, which carries nutrients between the roots and the leaves.

Across the culm wall, the number of fibres increases from the inside to the outside, as seen in Figure 3.16. There is also a variation in the shape of the fibres, with the outer fibres appearing almost circular, while the inner fibres are elliptical with a circular head facing the periphery. (Ray, et al., 2005).

Lignin supports the cellulose fibres and provides rigidity to the plant, making upright growth possible. It also improves the resistance of the plant against certain microorganisms (Janssen, 1981). The lignin in the cell walls is resistant to degradation, which helps to defend the plant against the invasion of pathogens and increases the durability of the plant (Yu, 2007).

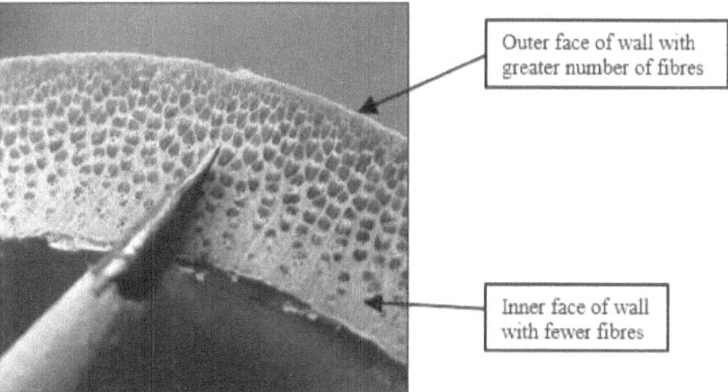

Figure 3.16: Close-up of a culm wall (Faville, 2012)

3.4.4   Vascular System

The *vascular system* or conducting tissue system is composed of *xylem* and *phloem* cells. The xylem cells transmit water and inorganic salts upwards, and the phloem cells transmit the products of photosynthesis (glucose and oxygen) downwards.

Within the nodes, at the diaphragms, the vascular bundles intersect, whereas within the internodes the bundles run parallel. At the same time, along the length of the walls, the conducting tissue of the vascular system increases from the outer part of the walls, with the maximum area in the middle third of the culm wall.

The organisation of the vascular bundles, the sclerenchyma cells, and the surrounding parenchyma cells are shown in Figure 3.17.

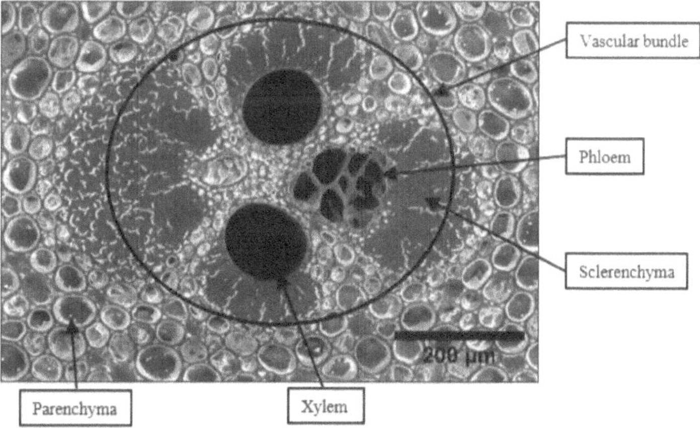

Figure 3.17: Scanning electron micrograph of a vascular bundle of sclerenchyma cells surrounded by the parenchyma matrix (Gerhardt, 2012)

3.4.5   Meristema

In bamboo plants, the dividing tissue is known as the *meristema* tissue. In bamboo, the diameter of the culm wall does not increase, so meristema tissue is only found in the internodes, where the cells increase the length of the culms (Chaowana, 2013).

3.5   Comparison of the Bamboo Plant with Timber

In the literature, bamboo has often been compared with timber, both as a structural material and as a construction technology. In order to make a comparison between timber and bamboo plants from a material point of view, it is necessary to identify the distinctive growth characteristics of timber.

A tree is made of up the following main components, namely the roots, the stem or trunk, and the crown (refer Figure 3.18). Water and nutrients are absorbed by the roots and transported to the leaves in the crown. The leaves manufacture food for the tree via the process of photosynthesis. This food is then transported to the areas of the tree where growth takes place, such as the tips of leaves and roots, and the cambium layer of the trunk.

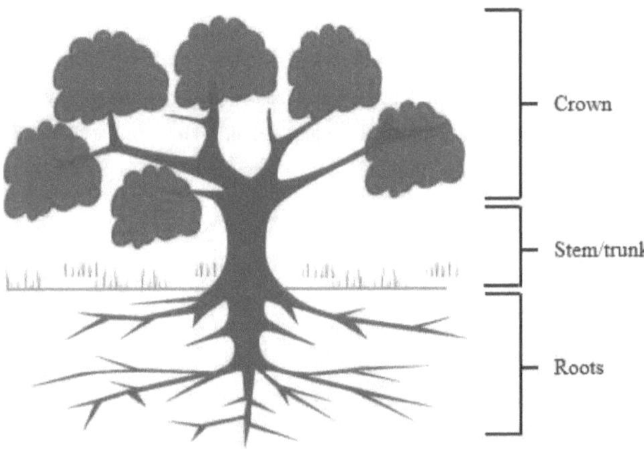

Figure 3.18: Main components of a tree (Max Pixel, n.d.)

In timber, the tissues in the trunk are formed as follows (Arbor Day Foundation, 2010) and shown in Figure 3.19:

1) *Outer bark* – the outer layers of the tree trunk
2) *Inner bark (phloem)* – the pipeline through which nutrients are passed to the rest of the tree
3) *Cambium* – the growing part of the trunk
4) *Sapwood* – new wood that moves water from the roots to the leaves
5) *Heartwood* – the central supporting pillar

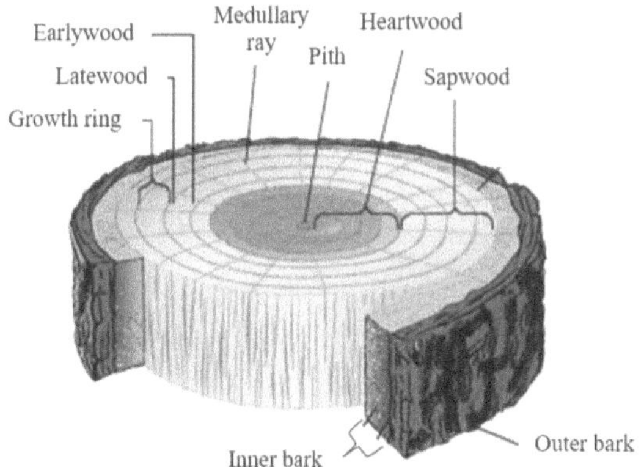

Figure 3.19: Layers found in a tree trunk (Trimmer, 2012)

3.5.1   Bark

The outermost layer of the tree trunk is called the bark of the tree. As the tree grows in circumference, the cells found in the inner bark die and become part of the outer layer. This layer is responsible for protecting the inner cells of the trunk. It keeps out moisture and prevents the inner cells from drying out.

3.5.2   Inner Bark

The inner bark, also known as phloem, conducts nutrients from the leaves to the growing portions of the tree. It only lives for a short period and becomes part of the outer bark when the cells die.

3.5.3   Cambium

The cambium is the layer of reproductive cells found between the inner bark and the sapwood and is responsible for the growth of the tree trunk in diameter. The cambium layer produces new sapwood cells towards the inside of the trunk, and new bark cells towards the outside of the trunk (Asif, 2009).

The growth that occurs during a single growing season is referred to as an annual or growth ring. Most trees make one growth ring per year.

3.5.4   Sapwood

The layer of the trunk just under the cambium layer is known as sapwood. This layer is the "living" part of the trunk and conducts water from the roots and stores the nutrients produced by the leaves (Asif, 2009).

3.5.5   Heartwood

Heartwood is the wood that extends from the pith (centre of the trunk) to the sapwood. It consists of dormant cells and provides strength to the tree. As the tree grows ages and increases in size, the inner layers of sapwood change to heartwood (Asif, 2009).

3.5.6 Shrinkage

As fresh-cut timber dries, it shrinks in size. Reports from the United Kingdom indicate a change of 1% in dimensional size for every 4% change in moisture content, with most of the shrinkage occurring across the grain (Trada, 2013).

3.5.7 Comparison

Bamboo is similar to wood due to its lignifying cell structure, whereby the plant cells become woody through the formation and deposit of lignin in cell walls. However, bamboo has a hard outer surface with a soft inner portion and exhibits no growth rings such as are found in typical wood plants. Typically, wood grows by increasing rings, while grasses such as bamboo grow by stretching. Grewal (2009) illustrated some of the anatomical differences between timber and bamboo, as seen in Table 3.1.

Table 3.1: Comparison between Bamboo and Timber (Grewal, 2009)

BAMBOO	TIMBER
One vascular system transporting both water (xylem) and nutrients (phloem)	Separate systems transporting water from the roots (inner bark) and nutrients from the leaves (sapwood)
Culms have the greatest strength at the outermost layer of the culm wall	The strongest wood is found at the centre (heartwood)
An outer layer of cutin and silica to protect the plant from insects and moisture	The outer bark provides protection from moisture and insects
Culm emerges with final diameter, increases in length with age, but not diameter	Tree trunk increases in diameter and circumference with time due to cambial growth (growth rings formed each year)
Develops as a linked underground network of roots, branching off to produce roots and stems	Each tree has its own root system
Reaches full maturity in 3 – 6 years	Reaches full maturity in 15 – 20 years or longer

3.6 Propagation & Growth

Bamboo is one of the fastest-growing plants, with an average daily growth rate of 20 cm, and can attain a growth rate of as much as 1 m per day for certain species. Within two to four months, bamboo plants can attain their full height of 15-30 metres, although they are not considered mature plants before 3 years of age. Diameters of culms vary from 10 mm to 300 mm, with wall thicknesses ranging from 5 mm to 50 mm. Growth rate and final size depend not only on the particular species but also on the locality and climate of the growth region.

When a new bamboo shoot emerges out of the ground, it is protected by a culm leaf sheath, which is later shed as the culm matures (refer Figure 3.20).

Figure 3.20: Bamboo shoots emerging from the ground (Schröder, 2010)

The nodes and internodes, with the associated diameter and length, are already determined in the new shoot (refer Figure 3.21), and do not change with age.

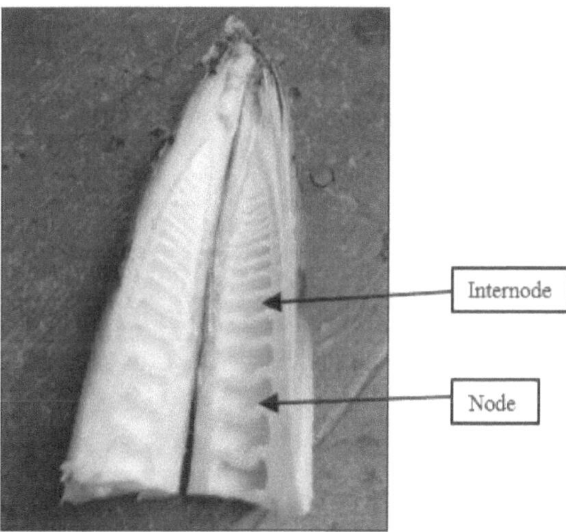

Figure 3.21: Cross-section of a bamboo shoot (Janssen, 2000)

Bamboo culms do not experience cambial growth like trees, which produce new tissue radially through the cambium, i.e. as the tree matures, so its diameter and circumference increases. Instead, bamboo culms emerge from the ground at their final diameter and can only create culms of larger diameter and height by the creation of new rhizomes with a larger diameter to the original. Each new culm generated from the rhizome emerges with a greater diameter and height than the previous culm. This process continues until the new culms reach the average size defined by the genetics of that species. Thus, a young clump of bamboo would consist of several culms with different diameters and heights, with the smallest located towards the middle of the clump, and the largest to the periphery (refer Figure 3.22) (Richard, 2013). Running bamboo would follow a similar pattern of increased growth in diameter, with the growth pattern being linear instead of circular.

After the initial period of growth, the culm matures, taking between three and five years, depending on the species. Bamboo does not experience any secondary growth. During the maturation process, lignification and silicification (hardening due to deposits of lignin and silicon in the cell walls) of the cell walls and parenchyma tissue continues as more lignin and silica accumulate. Studies vary on how long lignification continues, with some stating that the process is complete after one year of growth, while others state that the lignin content increases for up to seven years (Richard, 2013). The lignification process can also be affected by locality and climate, as these will dictate the quantity of starch that the culms are able to store. The cell wall thickening leads to a change in mechanical properties, with the result that some mechanical properties of bamboo increase with age, while others decrease. However, it has been reported that overall mechanical strength decreases after five years, as well as after flowering (Richard, 2013).

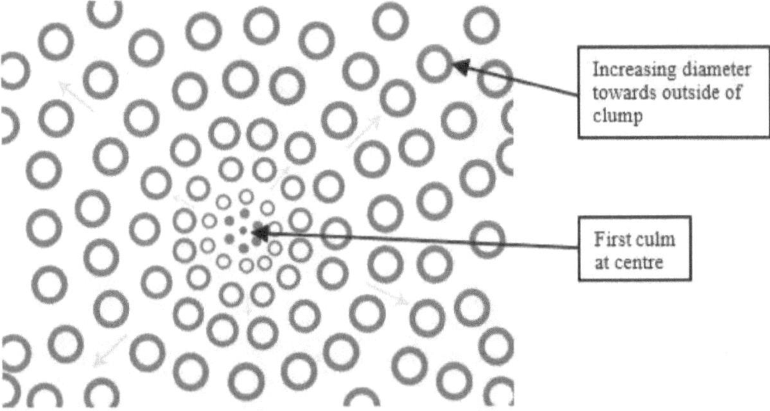

Figure 3.22: Formation of a bamboo clump (Schröder, 2012)

The culms remain free of branches until they have attained their full height. After this period of growth, the branches begin to grow from the nodal points. The bamboo plant sheds its leaves once a year (Laroque, 2007).

While most plants and grasses propagate by means of seeds, generated in the flowering season, bamboo is an exception. Bamboo plants flower very rarely, sometimes only once in a time period which varies from 15 to 100 years. Many bamboo plants flower only once in a lifetime and die thereafter. Other species flower annually but produce no fruit. The smaller species flower every three to four years, while the larger ones tend to flower every 20 to 120 years (refer Figure 3.23). Although the fruit that is produced is edible, it often falls to the ground before ripening. The seeds have a low ability to germinate, and so bamboo is usually propagated through means of cuttings (Janssen, 2000). Many species exhibit gregarious blooming, where all the members of a genotype or variety of a species, flower at once, irrespective of age or distribution. This blooming surge uses up the stored energy of the plants, leading to mass death of the bamboo culms. This flowering process takes place gradually, and it may be several years before death occurs.

Figure 3.23: Bamboo in flower (Shoumatoff, 2007)

The lifespan of an individual bamboo culm is approximately ten years, but this varies depending on the species. The maturity and lifespan of an individual culm are not necessarily the same as the maturity and lifetime of the bamboo plant. The plant is sustained by its root or rhizome system, so when an old culm dies, the plant continues to expand and generate new culms. The natural lifespan of a bamboo plant is approximately 50 years, but can be as long as 120 years, depending on the species.

3.7   Geographical Location

It has been estimated that there are more than 36 million hectares of bamboo plantations and forests worldwide. It occurs naturally in the tropical, subtropical and mild-temperate zones (46° N to 47° S latitude) (refer Figure 3.24), and from sea level up to 3800 m in elevation, with climates offering temperatures from 28°C to 50°C, although species also grow in colder areas, such as described in the following sections. The plants prefer well-drained soil, but also grow in wet and marshy locations. Thus, it can be surmised that bamboo grows in diverse physical circumstances. However, they do not tolerate saline soils or cold climates (Sharma, 2010).

There are 1662 different botanical species of bamboo (Canavan, et al., 2017), many of which have been used traditionally as structural members in footbridges, low-rise houses and roofs in countries where bamboo is a plentiful resource. Bamboo is reported to be one of the fastest-growing plants on earth, growing up to 1 m per day for some species, and has differing material characteristics and properties as a result of the different locations where it is grown.

Although bamboo originated in the tropics, they now cover a wide range of climates and altitudes. 14% of the bamboo species are known to have been introduced outside of their native ranges (Canavan, et al., 2017). INBAR has classified three major regions where bamboo can be found, namely the Asia-Pacific Region, the Americas Region and the African Region (Yuming & Chaomao, 2010), as shown in Figure 3.24.

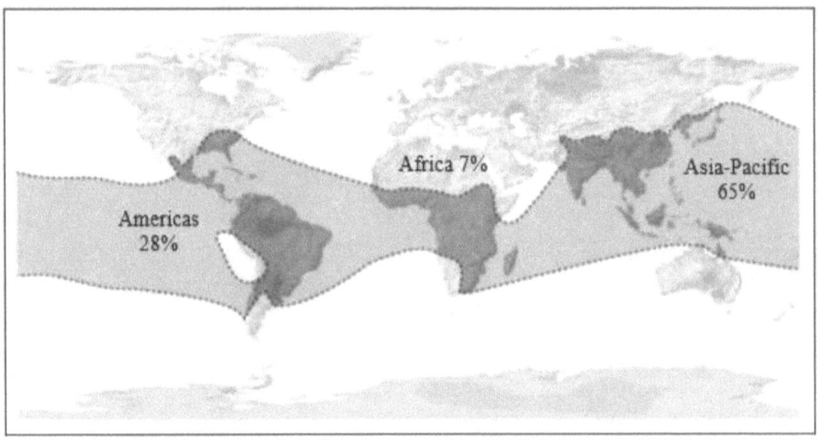

Figure 3.24: Global distribution of bamboo species (Sharma, 2010)

3.7.1 Asia-Pacific Region

The Asia-Pacific Region reaches from Singapore in the south in Asia, to Sakhalin Island in the north, the Pacific islands in the east, and the southwestern part of the Indian Ocean, with a distribution of more than 900 species of bamboo and 40 to 50 genera. The middle and southern parts of the Yunnan Provinces in China have the most abundant, naturally occurring bamboo species (Yuming & Chaomao, 2010).

Asia and Oceania produce 65% of the world's bamboo resources, with India producing the majority. A second large bamboo-producer is China, which has the greatest bamboo biodiversity in Asia, with over 500 species, mainly monopodial. However, no native bamboo is found north or northwest of Tibet and China.

- *India*

In India, more than 100 000 km^2 is occupied by bamboo species, constituting 12.8% of the total forest cover of the country (Bhatt, et al., 2004). The North Eastern Himalayan region has the highest diversity of bamboo, with 58 species out of the 126 species found in India, belonging to 18 genera. India has a typically warm humid climate with a mean annual rainfall of 2660 mm, most of which occurs during the south-west monsoon season (May to September). The maximum temperature ranges between 25 and 32 °C, with an average minimum of 11 °C, although the northern regions have been known to drop to -2°C in the winter months. The mean daily humidity varies from 80 - 96%, being highest during July to September, and lowest in March (Singh, et al., 2003).

Common Indian bamboo species used in construction are *Bambusa balcooa*, *Bambusa nutans*, *Dendrocalamus giganteus*, *Dendrocalamus strictus* and *Dendrocalamus asper* (Richard, 2013). Other bamboo species found in India are *Melocanna baccifera* or Muli bamboo, *Bambusa cacharensis*, *Dendrocalamus longispathus*

- *China*

In China, there are approximately 400-500 species in 40 genera (Yuming & Chaomao, 2010). China's State Forestry Administration reported a total bamboo forest area of 5.38 million hectares in 2009, most of which was found in plantations. In Southwest China, in the province of Yunnan, ten species of bamboo are widely cultivated or used. The average altitude is 1800 – 2000 meters, with mountain peaks at 4200 meters above sea level. The climate is subtropical monsoon with an average annual temperature of 12 – 17°C and an annual rainfall of 700 – 1000 mm (Long, et al., 2003).

The main bamboo species cultivated is Moso bamboo, or *Phyllostachys pubescens*, currently accounting for 80% of China's bamboo area (Yiping & Henley, 2010).

- *Hawaii*

Several species are found in Hawaii and are used for structural applications, namely *Bambusa beecheyana*, *Bambusa vulgaris*, *Guadua angustifolia*, *Dendrocalamus apus*, *Dendrocalamus strictus* (Bamboo Village Hawaii, 2004).

- *Malaysia*

In Malaysia, bamboo can be found in areas such as the foothills of the Main Range, forest fringes and river courses. The density of growth varies from place to place, depending on the site conditions, the extent of human disturbance in that area, and the particular species growing there. There are 50 known species of bamboo in Malaysia, of which 25 species are indigenous. The species *Gigantochloa scortechinii* is the most widespread and is most widely used by the inhabitants. Other indigenous species that are exploited are *Gigantochloa ligulata and Gigantochloa wrayi* (Wong, 1989).

- *Philippines*

In the Philippines, bamboo is grown in both private and public lands. In 2002, it was reported that there were approximately 13 530 hectares of government bamboo plantations within forestland, with 3040 hectares of private bamboo plantations (Pabuayon, 2004).

3.7.2 Americas Region

The Americas Region covers South America, Central and North America, from Chile in the south to the eastern region of the United States of America. It has approximately 270 species in 18 genera of woody bamboo plus more than 110 species of herbaceous bamboo and represents 28% of the world bamboo resources. Many of these species are of low economic value, apart from the genus *Guadua* (Yuming & Chaomao, 2010).

The Americas have the greatest diversity of bamboo, particularly South America, where countries such as Brazil have 134 naturally occurring bamboo species (Bystriakova, et al., 2004). Bamboo has a long history of use in construction in Central and South America, particularly in Colombia and Ecuador where Guadua angustifolia has been used to construct houses that have stood for 50 to 200 years.

- *Colombia, Ecuador and Venezuela*

The most common bamboo species in Colombia is *Guadua angustifolia*, covering 520 square kilometres, one of the species used most often for construction. It grows in altitudes up to 1800 m above MSL, in small forests and along creeks, with a diameter of approximately 180 mm. Two subspecies exist, namely *Guadua castilla* and *Guadua mecana* (Bystriakova, et al., 2004).

The favoured bamboo species in Ecuador is *Guadua angustifolia*, known for its physical and mechanical properties, as well as its large size. In Ecuador, it is known as 'caña guadúa', 'caña brava', 'caña macho' or sometimes just 'caña' (Klop, et al., 2003).

- *Costa Rica*

In Costa Rica, Central America, the National Bamboo Project uses two species of *Guadua* bamboo in house construction, namely *Guadua aculeate* and *Guadua angustifolia*.

3.7.3 African Region

The African Region stretches from Mozambique in the south to Sudan in the east and north, representing 7% of the total bamboo area in the world. Although information from Africa is incomplete, more than 2.7 million hectares of bamboo forest was reported in 2012 by six African countries (Zimbabwe, Tanzania, Uganda, Nigeria, Ethiopia and Kenya) (Bamboo Interim Steering Committee, 2012).

Bystriakova (2004) reported the lowest diversity of woody bamboo in the African region, with only thirty-eight naturally occurring species. Madagascar has the greatest diversity (thirty-eight species) while South Africa and Togo were each reported to have one. Overexploitation of the naturally growing bamboo has led to a decline in the natural stands. Some measures have been taken in Kenya and Uganda to counteract this whereby numerous species have been introduced. The most successful of these are *Bambusa bambos, Bambusa tulda, Bambusa vulgaris, Dendrocalamus asper, Dendrocalamus brandisii, Dendrocalamus membranaceus, Dendrocalamus strictus, Phyllostachys pubescens* and *Thyrsostachys siamensis.*

In terms of utilisation, the most significant African species are *Oreobambus buchwaldii, Oxytenanthera abyssinica* and *Arundinaria alpina.* Primary usage includes small-scale construction, handicrafts, fencing, water pipes, furniture and basketry (Bystriakova, et al., 2002).

- *Cameroon*

The Republic of Cameroon is located in Central Africa and covers a land area of 466 326 km^2. It has a tropical climate along the coast, becoming semi-arid to the north. The rainfall varies between 1500 mm in the south to 10 000 mm near Mount Cameroon (Ingram, et al., 2011). Cameroon has only two native bamboo species which are used for construction, namely *Oxytenanthera abyssinica* and *Yushania alpina*, also known as *Arundinaria alpina. Bambusa vulgaris* has also been reported as an introduced (non-indigenous) species in Cameroon.

- *Ethiopia*

Kibwage (2011) reported that there are over 850,000 hectares of lowland bamboo, *Oxytenanthera abyssinica*, and over 130,000 hectares of highland bamboo, *Arundinaria alpina*, in Ethiopia. Bamboo is a common building material for houses and fences as well as a source of food for cattle and people, as a result of a lack of alternative construction materials in the lowland areas.

- *Madagascar*

Madagascar has 33 species of woody bamboo, of which 32 are reported to be endemic. (Bystriakova, et al., 2004). The majority are found on the eastern escarpment where remnants of the rainforests still exist. Bamboo is used locally for construction, handicrafts and musical instruments, as well as water containers, fishing traps, and poles.

- *Mozambique*

The North and Central provinces of Mozambique have native bamboo in abundance. Duraisamy (2003) mentions that *Oxytenanthera* species is widespread in Africa to the extent that the genus is known as African bamboo (Duraisamy, 2003). This species is found in Kenya, Malawi, Mozambique, Uganda, Tanzania, Zambia and Zimbabwe. It is thin-walled, clumping bamboo, with holes in the upper nodes, and is often used for housing and granaries.

Three species of bamboo have been found in Malawi, namely *Arundinaria alpina, Oreobambus buchwaldii* and *Oxytenanthera abyssinica. Oxytenanthera abyssinica* is a medium-sized, semi-solid bamboo.

- *South Africa*

South Africa has only one indigenous species of bamboo, namely *Thamnocalamus tessellatus* or "bergbamboes" which is used to construct gates and screens, and used as walking sticks, implement handles and spears (Scheba, et al., 2017).

A second species, *Bambusa balcooa*, was purportedly introduced into South Africa by the Dutch East India Company in 1653. This species is considered to be a naturalised bamboo species.

There are several commercial bamboo initiatives which started in South Africa in the past ten years, such as the Kowie Farm owned by EcoPlanet Bamboo and the Green Grid Beema Bamboo Project in KwaZulu-Natal. This will be covered in more detail in Chapter 7.

3.7.4   Other areas

Although bamboo is typically not found in colder temperate regions such as Canada, Europe and the USSR, Fabiani (2014) reported that two bamboo species found in Italy, namely *Phyllostachys viridiglaucescens* and *Phyllostachys edulis*, would be suitable for the building of trusses and space frames (Fabiani, 2014).

3.8   Structural Bamboo Species

There are an estimated 121 genera of bamboo, comprising approximately 1662 species (Canavan, et al., 2017). The main species identified by INBAR as suitable for structural purposes are *Bambusa balcooa, Bambusa nutans, Bambusa polymorpha, Gigantochloa apus* and *Guadua angustifolia* (Jayanetti & Follett, 1998). Other common species used in construction include *Bambusa vulgaris, Phyllostachys pubescens, Bambusa stenostachya, Dendrocalamus strictus, Dendrocalamus asper* and *Dendrocalamus giganteus* (Richard, 2013).

The International Code Council (ICC) has evaluated bamboo poles from the species *Bambusa stenostachya*. It has deemed them suitable for use as structural elements in floors, walls, roofs and trusses, or as individual compression and tension members, in non-fire-resistance rated buildings, both commercial and residential construction (ICC Evaluation Service, 2013).

- *Bambusa balcooa*

Bambusa balcooa is non-invasive, forms clumps and has sterile seeds. It grows to an average height of 20 – 30 m with a culm diameter of 80 – 150 mm (refer Figure 3.25), with the cavity being approximately one-third of the culm and has a lifespan of approximately 300 years. It has nodes placed at 200 to 400 mm apart. It is reported as having gregarious flowering, with a flowering cycle of 35 to 45 years. The clumps die after flowering without setting any seed (INBAR, 2000). It can grow at any altitude up to 700 m, in any type of soil, but grows best in soil with good drainage.

Although it reaches maturity between five and eight years of age, it can be harvested at two to three years for construction purposes.

Figure 3.25: Bambusa balcooa (INBAR, 2000), (Schröder, 2014)

- *Bambusa nutans*

Bambusa nutans is a sympodial or clumping bamboo growing to 20 m in height, with strong straight culms often used as poles (Kalia, et al., 2004). It has a culm diameter of 50 – 100 mm (refer Figure 3.26), with a wall thickness of 25 mm. As well as being used in house construction and the paper industry, its smooth outer texture makes this species popular for basketry and craft products.

Figure 3.26: *Bambusa nutans* (Bamboo Land Nursery & Parklands, n.d.)

- *Bambusa polymorpha*

Bambusa polymorpha is found in North East India. It is a tall, greyish-green clumping bamboo with erect culms which curve outward at the top. It grows to 25 m in height, with a culm diameter of 70 - 150 mm and a wall thickness of 10 mm (refer Figure 3.27). It is often used for house construction and pulp for the paper industry. However, the natural durability of this species is low, often being destroyed within two years by termites and fungi if left untreated.

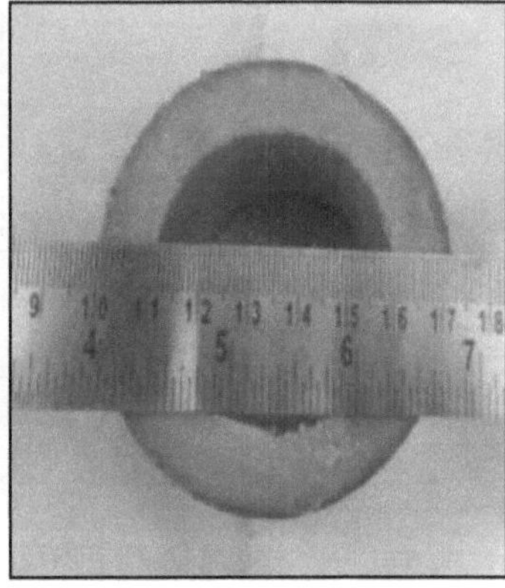

Figure 3.27: *Bambusa polymorpha* (Useful Tropical Plants Database, 2014), (Sanjib, 2018)

- *Bambusa stenostachya*

Bambusa stenostachya, also known as Tre Gai bamboo, is a clumping bamboo, with numerous branches at the node. The lower internodes are often almost solid, which makes it popular for construction purposes. It grows to 20 m in height, with a culm diameter of 150 mm and a wall thickness of 20 mm (refer Figure 3.28).

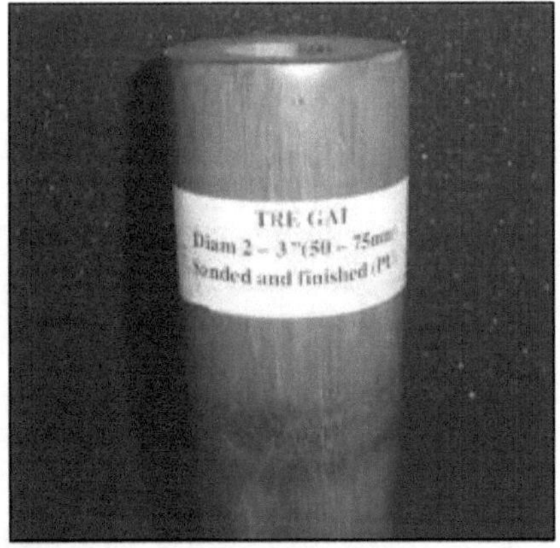

Figure 3.28: *Bambusa stenostachya* (BambooWeb.Info, n.d.)

- *Bambusa vulgaris*

Bambusa vulgaris is 10 - 35 m long, thick-walled with a lustrous green culm when young, 50 - 100 mm in diameter, with branches throughout the stem (refer Figure 3.29) (Hossain, et al., 2005). It has a high starch content, making it attractive to bore beetles. It is commonly used for fencing, buildings and boats.

Figure 3.29: ***Bambusa vulgaris*** **(Useful Tropical Plants Database, 2014)**

- *Dendrocalamus asper*

Dendrocalamus asper is also known as Giant bamboo or Dragon bamboo. It is a clumping bamboo with densely tufted culms and grows from 20 to 30 m in height, with a culm diameter of 80 – 200 mm and a wall thickness of 20 mm (refer Figure 3.30). The upper internodes of the culm are used as water containers and cooking pots. The culms are used as a building material for houses and bridges, as well as for furniture and musical instruments (refer Figure 3.31).

Figure 3.30: ***Dendrocalamus asper*** **(Bamboo Land Nursery & Parklands, n.d.)**

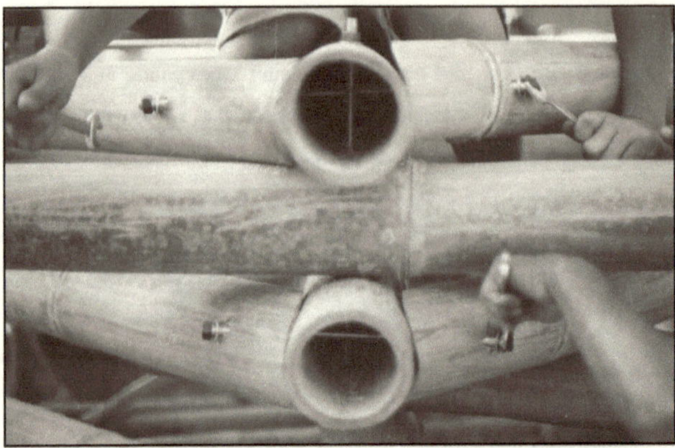

Figure 3.31: *Dendrocalamus asper* being used in construction (Fitrianto, 2014)

- *Dendrocalamus giganteus*

Dendrocalamus giganteus has been reported as having a daily growth rate of 100 to 300 mm, with a maximum of 580 mm per day (Nandy, et al., 2004). It is used for house construction and the production of boat masts. It ranks amongst the tallest bamboo at a maximum height of 30 m and a maximum diameter over 300 mm (refer Figure 3.32).

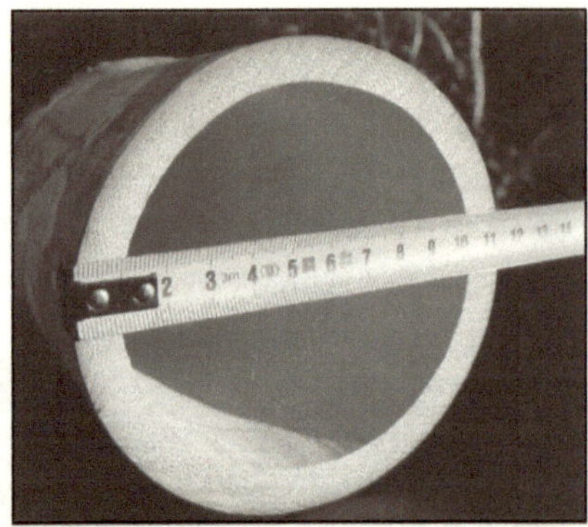

Figure 3.32: *Dendrocalamus giganteus* (Useful Tropical Plants Database, 2014)

- *Dendrocalamus strictus*

Dendrocalamus strictus, or Calcutta bamboo, is grown in 53% of the area in India where bamboo is grown. Although this species has been used for building in India, this has mainly been for temporary structures, and not permanent buildings (Liang & Robinson, 2009). It has a relatively small diameter of 25 to 80 mm, with a wall thickness between 15 mm and 75 mm (refer Figure 3.33). It grows from 8 to 20 m tall, but is curved above half its height; thus, only the lower portions are considered suitable for construction, with the culm lengths used limited to 5 m (Liang & Robinson, 2009). The intermodal length varies between 300 mm and 450 mm. It can grow at a rate of 370 mm per day (Grewal, 2009), reaching maturity at three years.

Figure 3.33: ***Dendrocalamus strictus*** **(Bamboo Land Nursery & Parklands, n.d.)**

- *Gigantochloa apus*

Gigantochloa apus is a dense clumping bamboo with short rhizomes. It grows from 8 to 30 m in height, with a culm diameter of 40 – 130 mm and a wall thickness of 15 mm (refer Figure 3.34). It is used for making cooking utensils, fishery utensils, furniture and baskets. It is also used as a building material for roofs, walls, scaffolding and bridges.

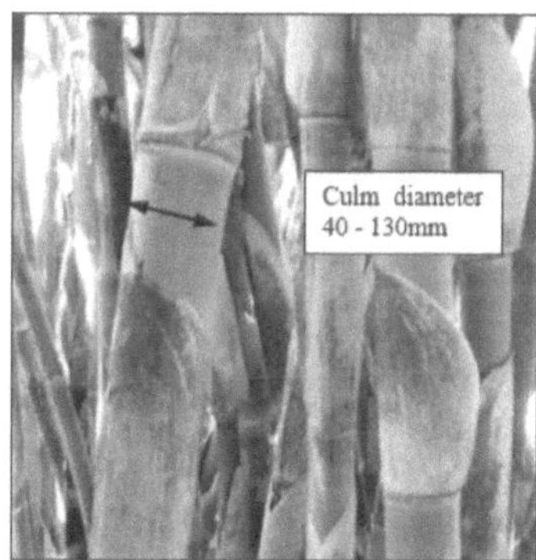

Figure 3.34: ***Gigantochloa apus*** **(Schröder, 2014),**

- *Guadua angustifolia*

Guadua angustifolia is a thorny, clumping bamboo. It grows up to 22 m in height, with a diameter of 120 mm and 15 mm wall thickness (refer Figure 3.35) (Quintans, 1998). While not having a large starch content, it is susceptible to beetle and fungal attack. It is widely used for house construction in Colombia and Ecuador.

Figure 3.35: *Guadua angustifolia* **(Bamboo Land Nursery & Parklands, n.d.)**

- *Phyllostachys pubescens*

Phyllostachys pubescens (previously known as *Phyllostachys edulis*) is also known as Moso bamboo or Tortoise-shell bamboo (refer Figure 3.36). It grows up to 20 m, with a diameter of 180 – 200 mm and 10 mm wall thickness (refer Figure 3.37). It is commonly used in the bamboo textile industry of China.

Figure 3.36: *Phyllostachys pubescens* **(BambooWeb.Info, n.d.)**

Figure 3.37: *Phyllostachys pubescens* **(BambooWeb.Info, n.d.)**

3.8.1 Physical Differences between species of Bamboo

As described here and in Chapter 4, the physical, chemical, material and strength properties vary between different species of bamboo. Culm dimensions, such as height, diameter and wall thickness, as well as age and material properties, determine the performance of bamboo in construction.

Much research has been done to tabulate the anatomical differences between various genera. Liese (1998) listed recent studies concerned with overall differences between species, their distinction and taxonomic relations. Furthermore, the vascular bundles vary in structure between the species, but this has been documented mainly for the Chinese bamboo species.

As mentioned already, on average, a culm consists of about 50% parenchyma, 40% fibres (sclerenchyma) and 10% conducting tissue (vascular system). However, these values vary between species and thus influence the material and structural properties of each species.

The different species of bamboo have different starch content, which affects their susceptibility to beetle attacks, as the beetles are attracted to the starch in the culms. Starch content is also affected by the season as well as the age of the bamboo, where the starch content is higher during the growing season and when the culms are young.

No recorded relationship has been found between the number of internodes and the height of the culm. Culms of the same clump and with the same height have differing numbers of internodes. Bamboo species with shorter internodes are preferred for structural load-bearing members, because the nodes provide stiffness to the culms, thus increasing the strength of the culm. Species with longer internodes are preferred when making containers. The internode length varies between species, from 200 mm to an extreme length of 4.5 m in species such as *Arthrostylidium schomburgkii* (Reubens, 2010).

Most bamboo species have straight, cylindrical culms, although some species, such as *Phyllostachys aureosulcata*, have serpentine bends. Some species, such as *Bambusa bambos* and *Guadua angustifolia*, develop crooked culms if grown in congested clumps. Some species have distinctive cross-sections, such as *Chimonobambusa quadrangularis*, also known as square bamboo, which has a square cross-section. Symmetry of the culms is also important when selecting species for structural applications. For example, *Phyllostachys pubescens*, which is popular in China, is almost perfectly straight and symmetrical, making it production friendly (Reubens, 2010).

The anatomical differences between these species of bamboo are summarised in Table 3.2.

Table 3.2: Physical differences between bamboo species

Species	Alternative names	Diameter	Wall Thickness	Height
Bambusa balcooa	Female bamboo, Giant bamboo	80 – 150 mm	25 – 50 mm	20 – 30 m

Table 3.2 (continued):

Species	Alternative names	Diameter	Wall Thickness	Height
Bambusa nutans	Burmese timber bamboo	50 – 100 mm	25 mm	20 m
Bambusa polymorpha	-	150 mm	10 mm	25 m
Bambusa stenostachya	Tre Gai bamboo, Thorny bamboo	150 mm	20 mm	20 m
Bambusa vulgaris	Common bamboo, Dragon head bamboo, Nepalese bamboo	50 – 100 mm	15 mm	10 – 35 m
Dendrocalamus asper	Giant bamboo	80 – 200 mm	20 mm	20 – 30 m
Dendrocalamus giganteus	Giant bamboo	300 mm	25 mm	25 – 60 m
Dendrocalamus strictus	Calcutta bamboo, Male bamboo, Solid bamboo	25 – 80 mm	15 – 75 mm	8 – 20 m
Gigantochloa apus	-	40 – 130 mm	15 mm	8 – 30 m
Guadua angustifolia	American narrow-leaved bamboo	120 mm	15 mm	30 m
Phyllostachys pubescens	Moso bamboo	180 – 200 mm	10 mm	10 – 20 m

3.9   Conclusions

Bamboo is the collective name for different species of giant grasses, belonging to the botanical family *Poaceae*. The sub-families are divided into tribes, which are further sub-divided into sub-tribes. Within each sub-tribe are a number of genera, which are further divided into individual species. Bamboo is found mainly in the tropical and subtropical regions. The main species found in each area have been documented.

Although there are more than 1660 different species of bamboo, few of these are suitable for construction purposes. Eleven species most commonly used for building construction were identified, and more anatomical detail was given on those species.

The macro- and microstructure of the bamboo culms are determined by their species, as well as their location. As only the culms are used for structural purposes, and not the roots, the microstructure of the culm walls was covered in detail. The macro- and microstructure also determine the material and engineering properties of the culms. This will be covered in more detail in Chapter 4 for the eleven species previously identified.

3.10 References

Arbor Day Foundation, 2010. *Anatomy of a Tree.* [Online] Available at: http://www.arborday.org/trees/treeguide/anatomy.cfm [Accessed 17 March 2017].

Asif, M., 2009. Sustainability of timber, wood and bamboo in construction. In: *Sustainability of Construction Materials.* Cambridge: Woodhead Publishing Limited, pp. 31-54.

Bamboo Botanicals, 2010. *Bamboo Anatomy and Growth Habits.* [Online] Available at: http://www.bamboobotanicals.ca/html/about-bamboo/bamboo-growth-habits.html [Accessed 18 March 2017].

Bamboo Interim Steering Committee, 2012. *Business Case for the National Bamboo Association of South Africa (NBASA),* Pretoria: s.n.

Bamboo Land Nursery & Parklands, n.d. *Bambusa nutans [image].* [Online] Available at: http://www.bambooland.com.au/assets/alt_2/BAMNUT.jpg [Accessed 30 March 2017].

Bamboo Land Nursery & Parklands, n.d. *Dendrocalamus asper [image].* [Online] Available at: http://www.bambooland.com.au/assets/thumbL/DENASPIND.jpg [Accessed 30 March 2017].

Bamboo Land Nursery & Parklands, n.d. *Dendrocalamus strictus [image].* [Online] Available at: http://www.bambooland.com.au/assets/alt_1/DENSTR.jpg [Accessed 30 March 2017].

Bamboo Land Nursery & Parklands, n.d. *Guadua angustifolia [image].* [Online] Available at: http://www.bambooland.com.au/assets/thumbL/GUAANG.jpg [Accessed 30 March 2017].

Bamboo Village Hawaii, 2004. *Bamboo Applications: Bamboo report - bamboo as a construction material.* [Online] Available at: http://www.bamboovillagehawaii.org/report.htm [Accessed 06 06 2011].

BambooWeb.Info, n.d. *Bambusa stenostachya [image].* [Online] Available at: http://www.bambooweb.info/images/bamboo/B.stenostachya.jpg [Accessed 21 March 2017].

BambooWeb.Info, n.d. *Phyllostachys pubescens [image].* [Online] Available at: http://www.bambooweb.info/images/bamboo/284/p1080368.jpg [Accessed March 21, 2017].

Bhatt, B. P., Singha, L. B., Sachan, M. S. & Singh, K., 2004. Commercial edible bamboo species of the North-Eastern Himalayan Region, India. Part 1: young shoot sales. *Journal of Bamboo and Rattan,* 3(4), pp. 337-364.

Bystriakova, N., Kapos, V. & Lysenko, I., 2002. *Potential distribution of woody bamboos in Africa and America,* Beijing: INBAR.

Bystriakova, N., Kapos, V. & Lysenko, I., 2004. *Bamboo biodiversity: Africa, Madagascar and the Americas.* [Online] Available at: http://www.unep-wcmc.org/resources/publications/UNEP_WCMC_bio_series/19.htm [Accessed 10 09 2011].

Canavan, S. et al., 2017. The global distribution of bamboos: assessing correlates of introduction and invasion. *AoB Plants,* Volume 9.

Chaowana, P., 2013. Bamboo: An Alternative Raw Material for Wood and Wood-based Composites. *Journal of Materials Science Research,* 15 March, 2(2), pp. 90-102.

Cronquist, A., 1988. *The Evolution and Classification of Flowering Plants.* 2nd ed. Bronx, New York: New York Botanical Garden.

Cyberscooty, n.d. *Parts of a bamboo plant [image].* [Online] Available at: http://clipart-work.net/data_gallery/bamboo-by-cyberscooty-bamboo-bambou-M3nrmD-clipart.png [Accessed 8 March 2017].

Duraisamy, J., 2003. Bamboo resources, enterprises and trade development opportunities for livelihood development and poverty reduction in Mozambique. *Journal of Bamboo and Rattan,* 2(4), pp. 429-439.

Fabiani, M., 2014. *Bamboo structures: Italian culms as likely resource for green buildings,* s.l.: Universita Politecnica delle Marche.

Faville, C., 2012. *The Idyll of the Split Bamboo.* [Online] Available at: http://compassrosebooks.blogspot.co.za/2012_10_01_archive.html [Accessed 2 October 2016].

Fitrianto, A., 2014. *Dendrocalamus [image}.* [Online] Available at: http://tropical.theferns.info/image.php?id=Dendrocalamus+asper [Accessed 15 September 2020].

Gerhardt, M. R., 2012. *Microstructure and mechanical properties of bamboo in compression,* Massachusetts: Massachusetts Institute of Technology.

Grewal, J. K., 2009. *Bamboo as a Structural Material,* Southampton: University of Southampton.

Hossain, M. A., Islam, M. S. & Hossain, M. M., 2005. Effect of light intensity and rooting hormone on propagation of Bambusa vulgaris Schrad ex Wendl. by branch cutting. *Journal of Bamboo and Rattan,* 4(3), p. 231241.

ICC Evaluation Service, 2013. *ESR-1636: Structural Bamboo Poles,* s.l.: ICC-ES.

INBAR, 2000. *Technical report 17: Bambusa balcooa.* [Online] Available at: http://www.inbar.int/publication/txt/tr17/bambusa/balcooa.htm [Accessed 27 02 2012].

Ingram, V. et al., 2011. *The bamboo production to consumption system in Cameroon,* Beijing: INBAR.

Janssen, J., 1981. *Bamboo in Building Structures,* Eindhoven: Eindhoven University of Technology.

Janssen, J., 2000. *Designing and building with bamboo,* Beijing: INBAR.

Jayanetti, L. & Follett, P., 1998. *Technical Report 16: Bamboo in Construction - An Introduction,* Beijing: INBAR.

Kalia, S., Kalia, R. K. & Sharma, S. K., 2004. In vitro regeneration of an indigenous bamboo (Bambusa nutans) from internode and leaf explant. *Journal of Bamboo and Rattan,* 3(3), pp. 217-228.

Kelchner, S., 2013. Higher level phylogenetic relationships within the bamboos (Poaceae: Bambusoideae) based on five plastid markers. *Molecular Phylogenetics and Evolution,* 67(2), pp. 404-413.

Kellogg, E. A., 2001. *Evolutionary History of the Grasses,* St Louis: American Society of Plant Biologists.

Kibwage, J. K., Frith, O. B. & Paudel, S. K., 2011. *Bamboo as a building material for meeting East Africa's housing needs: a value chain case study from Ethiopia,* Beijing: INBAR.

Klop, A., Cardenas, E. & Marlin, C., 2003. Bamboo production chain in Ecuador. *Journal of Bamboo and Rattan,* 2(4), p. 327343.

Kumar, S., 2017. *Internal Structure of Leaf (With Diagram).* [Online] Available at: http://www.biologydiscussion.com/leaf/internal-structure-leaf/internal-structure-of-leaf-with-diagram/14002 [Accessed 18 March 2017].

Laroque, P., 2007. *Design of a Low-Cost Bamboo Footbridge,* Massachusetts: Massachusetts Institute of Technology.

Liang, C. B. & Robinson, P., 2009. *Bamboo as a Permanent Structural Component,* London: Imperial College London.

Liese, W., 1998. *The Anatomy of Bamboo culms,* Beijing: INBAR.

Long, C., Liu, Y., Xue, J. & Li, D., 2003. A bamboo germplasm collection for community development in Central Yunnan, China. *Journal of Bamboo and Rattan,* 2(1), pp. 3-11.

Mannan, S., Knox, J. P. & Basu, S., 2017. Correlations between axial stiffness and microstructure of a species of bamboo. *Royal Society Open Science,* 4(1).

Max Pixel, n.d. *Tree [image].* [Online] Available at: http://maxpixel.freegreatpicture.com/static/photo/640/Nature-Plant-Root-Ecology-Tree-Branches-Eco-309046.png [Accessed 19 March 2017].

Nandy, S., Das, A. K. & Das, G., 2004. Phenology and culm growth of Melocanna baccifera (Roxb.) Kurtz in Barak Valley, North-East India. *Journal of Bamboo and Rattan,* 3(1), pp. 27-34.

Pabuayon, I., 2004. Consumption value and income potential of bamboo in the Philippines: Evidence from selected areas and case studies. *Journal of Bamboo and Rattan,* Volume 3, pp. 249-263.

Quintans, K. N., 1998. *Ancient Grass, Future Natural Resource,* Beijing: INBAR.

Ray, A. K., Mondal, S., Das, S. K. & Ramachandrarao, P., 2005. Bamboo - A functionally graded composite-correlation between microstructure and mechanical strength. *Journal of Materials Science,* Volume 40, pp. 5249-5253.

Reubens, R., 2010. *Bamboo in Sustainable Contemporary Design,* Beijing: INBAR.

Richard, M. J., 2013. *Assessing the Performance of Bamboo Structural Components,* Pittsburgh: University of Pittsburgh.

Sanjib, B., 2018. *Comparative Studies of physico chemical properties of selected less utilised Bamboo species of Kokrajhar District of BTAD Assam India,* Assam, India: Bodoland University.

Scheba, A., Blanchard, R. & Mayeki, S., 2017. *Bamboo for green development? The opportunities and challenges of commercialising bamboo in South Africa,* Cape Town: HSRC.

Schröder, S., 2010. *Bamboo shoot [image].* [Online] Available at: https://www.guaduabamboo.com/species/dendrocalamus-latiflorus [Accessed 19 March 2017].

Schröder, S., 2012. *Formation of a bamboo clump [image].* [Online] Available at: https://www.guaduabamboo.com/guadua/guadua-bamboo-growing-habits [Accessed 19 March 2017].

Schröder, S., 2014. *Bambusa balcooa [image].* [Online] Available at: https://www.guaduabamboo.com/blog/bambusa-balcooa [Accessed 15 September 2020].

Schröder, S., 2014. *Gigantochloa apus [image].* [Online] Available at: https://www.guaduabamboo.com/species/gigantochloa-apus [Accessed 25 March 2017].

Shamburg, M., 2010. *The making of a bamboo spey rod.* [Online] Available at: http://www.speypages.com/speyclave/59-rods/42442-making-bamboo-spey-rod.html [Accessed 02 October 2016].

Sharma, B., 2010. *Seismic Performance of Bamboo Structures,* Pittsburgh: University of Pittsburgh.

Shelton, P., 2009. *Bamboo Plants Part 1.* [Online] Available at: http://www.halfsidebamboo.info/wordpress/96/bamboo-plant-parts-i/ [Accessed 20 March 2017].

Shoumatoff, A., 2007. *Bamboo in flower [image].* [Online] Available at: http://www.vanityfair.com/news/2007/12/famine200712 [Accessed 19 March 2017].

Singh, H. B., Kumar, B. & Singh, R. S., 2003. Bamboo resources of Manipur: an overview for management and conservation. *Journal of Bamboo and Rattan,* 2(1), pp. 43-55.

Trada, 2013. *WIS 0-5 Timber frame building: materials specification,* Buckinghamshire: Trada Technology.

Trimmer, J., 2012. *Tree diagram [image].* [Online] Available at: http://www.woodstairs.com/tradepage/wp-content/uploads/2012/07/Tree-Diagram.png [Accessed 20 March 2017].

Useful Tropical Plants Database, 2014. *Bambusa polymorpha [image].* [Online] Available at: http://tropical.theferns.info/viewtropical.php?id=Bambusa+polymorpha [Accessed 21 March 2017].

Useful Tropical Plants Database, 2014. *Bambusa vulgaris [image].* [Online] Available at: http://tropical.theferns.info/viewtropical.php?id=Bambusa+vulgaris [Accessed 21 March 2017].

Useful Tropical Plants Database, 2014. *Dendrocalamus giganteus [image].* [Online] Available at: http://tropical.theferns.info/viewtropical.php?id=Dendrocalamus+giganteus [Accessed 21 March 2017].

Wong, K. M., 1989. Current and Potential Uses of Bamboo in Peninsular Malaysia. *Journal of the American Bamboo Society,* 7(1), pp. 1-15.

Yiping, L. & Henley, G., 2010. *Biodiversity in Bamboo Forests: A Policy Perspective for Long Term Sustainability.,* Beijing: INBAR.

Yuming, Y. & Chaomao, H., 2010. *China's Bamboo,* Beijing: INBAR.

Yu, X., 2007. *Bamboo: Structure and Culture,* s.l.: University of Duisburg-Essen.

Chapter 4

4 BAMBOO AS A CONSTRUCTION MATERIAL

4.1 Introduction

Each bamboo species has its own particular anatomical structure, which determines the mechanical and structural properties thereof. These anatomical structures and variations were addressed in Chapter 3.

This chapter will address the variations found in the physical and mechanical properties of bamboo culms, as derived from the anatomical properties. Furthermore, the effect of these variations in properties on the design process will be covered briefly, with more detail on design in Chapter 5. The main species covered will be the species recommended for construction purposes, as highlighted in Chapter 3. These eleven species are as follows: *Bambusa balcooa, Bambusa nutans, Bambusa polymorpha, Bambusa stenostachya, Bambusa vulgaris, Dendrocalamus asper, Dendrocalamus giganteus, Dendrocalamus strictus, Gigantochloa apus, Guadua angustifolia* and *Phyllostachys pubescens.*

The properties for these species are summarised under each section. A full summary of the physical and mechanical properties, as well as other documented results, is given in Appendix A.

4.2 Factors affecting properties

Every type of construction material has its own unique properties, based on the composition of the material itself. In this book, these properties considered are *physical* properties and include density, Elastic Modulus and Poisson's ratio. Furthermore, each type of material has *movement* properties, where the material experiences movement or dimensional change due to the reaction of the physical components of the material with the environment. These properties include shrinkage and creep. Lastly, each type of material has *mechanical* properties, which are derived from the physical properties. These properties determine how the material can be used in a structure, and include compressive and tensile capacity, as well as shear and flexural capacities.

Bamboo is both a non-homogeneous and an anisotropic material, with differing composition, structure and character throughout the length and position in the culm, as discussed in Chapter 3. These factors affect the physical and mechanical properties of the culms in the longitudinal, radial and tangential directions. The main factors affecting these properties are summarised in Table 4.1, with reference to the sections under which they are discussed further.

The physical and mechanical properties of a bamboo culm are affected by the anatomical properties. For example, the mechanical properties are determined by the density thereof, which in turn is largely dependent on the cell wall thickness, diameter and fibre content of the culm. Shrinkage of the culm is also affected by the stage of fibre maturation, with the older culms exhibiting less shrinkage than younger culms. The fibre length also affects the Modulus of Elasticity and compressive strength (Liese, 1998).

Table 4.1: Factors affecting Properties of Bamboo culms

Factor	Property affected	Section Reference
Variation in number of fibres in the culm wall	Density	4.3.1
	Poisson's ratio	4.3.3
	Creep and deformation	4.3.5
Variation in cross-section along length of culm	Density	4.3.1
	Elastic Modulus	4.3.2
	Shrinkage	4.3.4
	Creep and deformation	4.3.5
	Tensile strength	4.4.1
Moisture content	Elastic Modulus	4.3.2
	Compressive strength	4.4.1
	Bending strength	4.4.3
	Shear strength	4.4.4
	Shrinkage	4.3.4
	Creep and deformation	4.3.5
Age of culm	Shrinkage	4.3.4
	Creep and deformation	4.3.5
Environmental growth conditions	Elastic Modulus	4.3.2
	Poisson's ratio	4.3.3
	Compressive strength	4.4.1
	Tensile strength	4.4.1

4.3 Physical and Deformation Properties

4.3.1 Density

The density of bamboo ranges between $400 - 900$ kg/m^3, depending on the anatomical structure and position in the culm, with the highest density occurring at the outermost portion of the wall, and at the base of the culm (Chaowana, 2013). The values found in the literature vary regarding species, as well as with source of information. Janssen (2000) reported the average density of most bamboos being 700-800 kg/m^3, varying with the location of where the bamboo is grown, the species and the position in the culm, while Xiao (2010) reported the average density of bamboo culms as 650-800 kg/m^3.

Actual reported values for the various species are given in Table 4.2, with differentiation between dried or green specimens given where available. It was unclear in most cases which portion of the culms was tested, whether nodes were included in the specimens, or the position along the length of the culms (i.e. top, middle or base). This lack of clarity could explain the variability in the values between the species.

Table 4.2: Density of bamboo species

Bamboo species	Density (kg/m³)	Reference
Bambusa balcooa	783 (green)	(Bureau of Indian Standards, 2015)
	720 - 850 (dried)	(Naik, 2005)
Bambusa nutans	673 (dried)	(Bureau of Indian Standards, 2015)
	810 - 990 (dried)	(Naik, 2005)
Bambusa polymorpha	540 - 790 (dried)	(Jansen & Duriyaprapan, 1995)
Bambusa stenostachya	673 (dried)	(ICC Evaluation Service, 2017)
Bambusa vulgaris	510 - 720 (dried)	(Mbuge, 2000)
	626 (green)	(Bureau of Indian Standards, 2015)
Dendrocalamus asper	700 - 800	(Brink, 2008)
Dendrocalamus giganteus	700 - 810 (dried)	(Naik, 2005)
Dendrocalamus strictus	500 - 800	(Liang & Robinson, 2009)
	540 – 780 (dried)	(Jansen & Duriyaprapan, 1995)
	728 (dried)	(Bureau of Indian Standards, 2015)
Gigantochloa apus	202 - 415 (dried)	(Nugroho, et al., 2018)
Guadua angustifolia	400 – 800 (dried)	(Aijazi, 2013)
Phyllostachys pubescens	700	(van der Lugt & Vogtländer, 2015)
	677 - 788 (dried)	(Lorenzo, et al., 2019)

The range of density of bamboo for most species is similar to timber which varies from 500 - 700 kg/m³ (SANS 10160-2, 2011) and is substantially lighter than steel at 7800 kg/m³. This lightness can be both an advantage and a disadvantage, depending on the circumstances.

The lightness of this material has advantages in conventional construction such as reducing the load on the foundations; thus, the foundations can be smaller and less costly. Also, in a seismic situation, the low mass will result in a reduction in the seismic loads on the structure, which is advantageous.

In addition, this light weight combined with its high ductility makes it advantageous for use in structures required to withstand seismic forces. The light weight reduces the seismic loads to be resisted, and the flexibility and ductility enable the structure to absorb the forces with less damage than a more rigid material.

However, in high wind situations, it is desirable to have more weight in the building to counteract the wind forces. In areas of relatively high wind forces, heavier foundations or sturdier structures would improve the stability of the structure and assist in counteracting the wind forces, with the bamboo culms connected to the foundations in such a way as to allow them to function as a unit, and therefore take advantage of the heavier foundations.

4.3.2 Elastic Modulus

The Elastic Modulus (also called Young's Modulus or Modulus of Elasticity, MOE) of a material can be defined as the slope of its stress-strain curve in the elastic deformation region, or the ability of a material to resist deformation when a stress is applied to it.

Pawar (2014) stated that bamboo has a Modulus of Elasticity of 15 to 20 GPa, but no indication was given of the source or species of these figures. Janssen (2000) stated that the MOE for bamboo could be calculated theoretically as 350 x percentage of fibres in the sample. Thus, due to the variance of the fibres in the inner and outer walls of the culm, so too the MOE values would vary, namely 3.5 GPa for the inner wall and 21 GPa for the outer wall.

Actual reported values for the various species are given in Table 4.3. It is unclear in most cases as to whether these results refer to green or dried bamboo, their position in the culm or which part of the culm was tested. This lack of clarity could explain the variation in the values between the species.

Table 4.3: Elastic Modulus values for bamboo species

Bamboo species	Elastic Modulus (GPa)	Reference
Bambusa balcooa	7.3 (green)	(Bureau of Indian Standards, 2015)
	9.3 - 12.7 (dried)	(INBAR, 2000)
	9.6 - 16.9 (dried)	(Naik, 2005)
	10.9	(Anokye, et al., 2016)
Bambusa nutans	10.7 (dried)	(Bureau of Indian Standards, 2015)
	12.9	(Anokye, et al., 2016)
	20.0 - 21.5 (dried)	(Naik, 2005)
Bambusa polymorpha	7.0 - 9.6 (dried)	(Jansen & Duriyaprapan, 1995)
Bambusa stenostachya	11.1 - 16.5 (dried)	(Dixon, et al., 2015)
	11.7 (dried)	(ICC Evaluation Service, 2017)
	13.5 (dried)	(Richard, 2013)
Bambusa vulgaris	2.9 (green)	(Bureau of Indian Standards, 2015)
	6.1	(Chaowana, 2013)
	6.1 - 8.8	(Widjaja & Risyad, 1980)
	6.9 - 12.4 (dried)	(Mbuge, 2000)

Table 4.3 (continued):

Bamboo species	Elastic Modulus (GPa)	Reference
Dendrocalamus asper	6.3	(Chaowana, 2013)
	12.2 - 14.9	(Widjaja & Risyad, 1980)
Dendrocalamus giganteus	12.2 - 17.2	(Widjaja & Risyad, 1980)
	14	(Baba & Bhalla, 2009)
	14.2 - 20.0 (dried)	(Naik, 2005)
Dendrocalamus strictus	6 – 11.8 (dried)	(Jansen & Duriyaprapan, 1995)
	12.7 - 17.8	(Ahmad & Kamke, 2005)
	15	(Liang & Robinson, 2009)
	15 (dried)	(Bureau of Indian Standards, 2015)
	15.9	(Anokye, et al., 2016)
Gigantochloa apus	10.9 - 25.4 (dried)	(Nugroho, et al., 2018)
Guadua angustifolia	14.8 - 24.6 (dried)	(Dixon, et al., 2015)
	17.6 (dried)	(Gnanaharan, 1994)
	19	(Yu, 2007)
Phyllostachys pubescens	10.5	(Yu, 2007)
	10.8 - 13.9	(Lorenzo, et al., 2019)
	12.6 - 17.2 (dried)	(Dixon, et al., 2015)

4.3.3 Poisson's Ratio

Janssen (1981) determined Poisson's ratio for bamboo from tensile tests, with an average value of 0.51 for the outer skin of the culm wall and 0.31 for both the wall thickness and the inner skin. The higher value for the outer skin could be due to the presence of the silica layer on the outer face of the culm, and thus should not be considered representative of the value for the culm wall. Janssen cited a value of 0.32 obtained by Cox and Gaymayer in 1969, which correlates closely with Janssen's results, and his theoretical value of 0.3 for lignin. There was no mention made of the species tested, whether the bamboo culms were green or dried, or which portion of the culms were used in testing.

Sharma (2010) cited a study done in 2002 on the species *Guadua angustifolia* that gave an average Poisson's ratio of 0.27 with a standard deviation of 0.07.

No further information was found in the literature regarding Poisson's Ratio for bamboo.

4.3.4 Moisture content and Shrinkage

The mechanical and physico-chemical properties of green bamboo are affected by the moisture content of the culm as well as by the cellulose-lignin content.

The moisture content of bamboo depends on species, position in the culm, season and age of the culm (Yu, 2007). Different species have differing numbers of parenchyma cells (soft cell tissue used to store food and nutrients) which determine the water-holding capacity of the culm. The top of the culm is reported to have a lower moisture content than the base; in addition, the inner part of the culm has higher moisture content than the outer portion. Culms have higher moisture content at the end of the rainy season than during or at the end of the dry season. Also, younger culms have a higher moisture content than mature culms. After harvesting, the moisture content can be influenced by the humidity and dryness of the environment.

Jiang (2012) tested *Phyllostachys pubescens (*Moso) bamboo samples for sensitivity to moisture content. His results indicated that compression and shear strength were the most sensitive, followed by longitudinal tensile strength and bending modulus. He stated that a change of 1% in moisture content would change the compressive strength by 3.8%, shear strength by 3.1%, tensile strength by 1.6% and bending modulus by 1.5%.

Reported values for moisture content are given in Table 4.4, where available. The National Building Code of India states that mature culms of any species should have a moisture content of 20%, while the ICC (2017) recommends values less than 16%, both of which are higher than the reported values in Table 4.4.

Most of the literature provided only single values for the moisture content of the various species, without mention of the process followed to obtain those values. It has been assumed that the values given are average moisture content values for each species, and do not indicate an absolute value. As indicated in Chapter 3, each species has different material composition and characteristics, which are also affected by the climate and location where a particular species is grown. Thus, these reported moisture content values should be considered solely as a theoretical guideline.

Table 4.4: Moisture content in Bamboo species

Bamboo species	**Average Moisture content (%)**	**Reference**
Bambusa balcooa	8.5	(Naik, 2005)
Bambusa nutans	8.0	(Naik, 2005)
Bambusa polymorpha	n.a.	-
Bambusa stenostachya	n.a.	-
Bambusa vulgaris	9.9 - 11.9 (dried)	(Mbuge, 2000)
Dendrocalamus asper	n.a.	-
Dendrocalamus giganteus	8.0	(Naik, 2005)
Dendrocalamus strictus	9 - 10	(Liang & Robinson, 2009)
Dendrocalamus strictus	4.2 - 9.5	(Ahmad & Kamke, 2005)
Guadua angustifolia	15	(Richard, 2013)

Table 4.4 (continued):

Bamboo species	Average Moisture content (%)	Reference
Gigantochloa apus	15.1 - 18.2 (dried)	(Nugroho, et al., 2018)
Phyllostachys pubescens	9.4 - 13.4	(Lorenzo, et al., 2019)

Note: n.a. = not available

Like many materials such as timber, bamboo culms shrink when they lose water, which can lead to cracks and splitting (refer Figure 4.1).

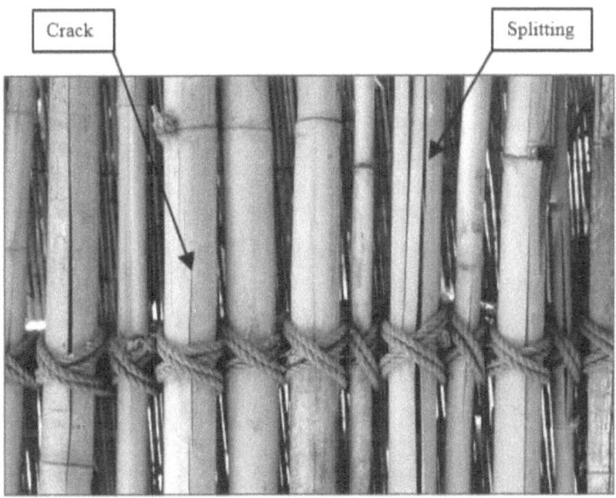

Figure 4.1: Cracks in bamboo culms (Anon., n.d.)

On average, the diameter of a bamboo culm reduces by 10-16% during shrinkage, while the wall thickness reduces approximately 15-17% (Burger, et al., 2017). Shrinkage is influenced by the stage of fibre maturation and the density of vascular bundles. Thus, older culms will exhibit less shrinkage, and less shrinkage will occur at the top of the culm than at the base due to the higher number of vascular bundles at the top (Liese, 1998). In addition, the moisture content is higher in the base of the culm, as mentioned previously. The dimensional shrinkage varies in different orthotropic directions, with radial shrinkage being the greatest and longitudinal shrinkage the least (Chaowana, 2013).

4.3.5 Creep and Deformation

Creep can be defined as "the inelastic deformation of a material due to changes in the material caused by the prolonged application of stress" (Gottron, 2014), where the creep characteristics depend on the loading conditions as well as the material being loaded. Little research was found regarding the creep behaviour of bamboo.

Janssen stated that flexural creep in bamboo was considered negligible due to the insignificant effects of the sustained loads, with permanent plastic deformation being 3 – 5% of the immediate elastic deformation (Janssen, 2000). This understanding is similar to the recommendation of the International Code Council (ICC), who recommended that long-term flexural creep in bamboo in bending should be assumed to be 3 – 4% of the immediate elastic deformation for the species *Bambusa stenostachya* (ICC Evaluation Service, 2017).

Although creep behaviour of bamboo may occur with loading along the axis of the culm, loads perpendicular to the culm often dictate the failure of the bamboo culms (Gottron, 2014). This is covered in more detail in Section 4.5.

Gottron studied the creep behaviour of *Bambusa stenostachya* bamboo specimens under four-point bending. The specimens were cut radially from the culms and tested for moisture content and ultimate load capacity (refer Figure 4.2) prior to creep testing. She compared the results from her tests with the acceptance criteria for wood as given by ASTM D6815-09 *Standard Specification for Evaluation of Duration of Load and Creep Effects of Wood and Wood-Based Products* and concluded that the relative creep of the bamboo specimens met or exceeded the criteria for wood. Contrary to Janssen (1981), Gottron concluded that creep-induced plastic deformation of bamboo was significant and should be considered similarly to creep in wood.

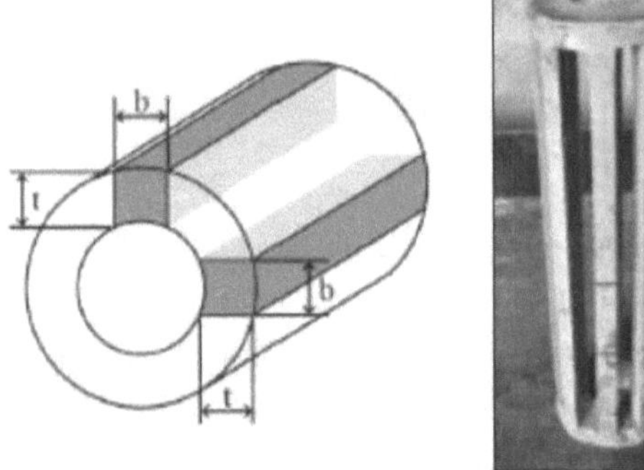

Figure 4.2: Test specimen machining (Gottron, et al., 2014)

Gottron (2014) also reported that the density of the bamboo culm had a more significant effect on the initial deformation and a negligible effect on the long-term creep. She stated that denser bamboo was stiffer, thus exhibiting lower elastic deformation than less dense bamboo.

Insufficient information was found to determine whether flexural creep in bamboo is significant or not, as the tests gave conflicting results. In addition, the tests reported on used samples of bamboo cut from the culm walls and were not conducted on full culm specimens. Further testing is required on the issue of creep in full culm samples to determine whether this could significantly affect structures.

4.3.6   Thermal performance

Thermal properties of building materials affect the environmental performance of the structure or building by dictating the energy required for heating and cooling of the various materials, and thus, in effect, the heating and cooling of the building (Shah, et al., 2016).

Huang et al. (2014) evaluated samples cut from the culm wall of *Phyllostachys pubescens* (previously known as *Phyllostachys edulis*) for thermal conductivity and diffusivity. Their test results were inconclusive, and they attributed this to the non-homogenous nature of bamboo culms.

Apart from the two articles as mentioned above, no information was found regarding the thermal performance of bamboo, indicating that this aspect of bamboo construction needs further study.

In addition, the irregularity of most bamboo culms would need to be considered when reviewing the thermal performance of the building envelope, and not only the materials making up the building or structure. The non-uniform shape of most bamboo culms could result in gaps that would need to be sealed to achieve the full thermal potential of the material.

4.3.7   Behaviour in Fire

Fires that occur in buildings and structures are sometimes deliberate, and sometimes accidental. One aspect of the design of buildings and structures is to ensure that these structures are adequately designed and built to manage potential fires.

The International Building Code (2015) states that buildings and their components should be designed to "safeguard against the spread of fire and smoke within a building, as well as the spread of fire to or from buildings". The National Building Code of India (2015) elaborates further, stating that buildings should contribute to the safety of life from fire, fumes, smoke and panic arising from fires.

The South African Building Regulations, SANS 10400-T (2011), provides more detail, stating that buildings should be designed and constructed so that in the event of fire:

a) occupants or users can evacuate safely,

b) the intensity and spread of such a fire is minimised,

c) the building stability is retained such that no other building is endangered,

d) the generation and spread of smoke are minimised or controlled,

e) there is adequate access and equipment for detecting, fighting, controlling and extinguishing such a fire (SANS 10400-T, 2011).

In order to meet the above criteria, it is important to consider how the building materials will behave when exposed to fire. This behaviour is assessed by considering two aspects: (1) fire reaction, and (2) fire resistance.

Fire reaction describes how easily or quickly a material ignites and spreads fire. In contrast, *fire resistance* is defined as the time that an element can maintain its structural integrity and stability in the presence of fire (Mena, et al., 2012). This length of time is often referred to as the material's *fire resistance rating* (FRR) (Kaminski, et al., 2016).

The International Building Code (2015) classifies the fire resistance of buildings and structures either by type of construction, or by use and occupancy. The type of construction identifies the type of materials used in the construction of the building and classifies the fire resistance and the level of combustibility associated with the construction elements. The term 'combustibility' refers to the tendency of a substance to burn as a result of fire or chemical reaction and is a measure of how easily a substance will burn or ignite (Bureau of Indian Standards, 2015).

The International Building Code has five construction types range from I to V, where Type I construction requires the highest fire-resistance rating and Type V requires the least. These are defined as follows (IBC, 2015):

- Type I and II: Building elements (excluding walls) are made from non-combustible materials, where Type II buildings are made from elements with a lower fire rating than Type I.
- Type III: Exterior walls are made from non-combustible material, while the interior elements are made from any material.
- Type IV: Exterior walls are made from non-combustible material, while the interior elements are either made from or incorporate timber.
- Type V: Structural elements are made from any permissible materials (IBC, 2015).

The National Building Code of India follows a similar approach to the International Building Code but has only four construction types (refer Table 4.5).

Table 4.5: Fire resistance rating of various building elements (Bureau of Indian Standards, 2015), (IBC, 2015)

	Fire resistance rating (minutes)				
Structural element	**Type I**	**Type II**	**Type III**	**Type IV**	**Type V**
Exterior load-bearing walls:					
International Building Code	180	60	120	120	60
National Building Code of India	240	120	120	60	-
Primary structural frame:					
International Building Code	180	60	60	60	60
National Building Code of India	180	90	60	60	-
Roof construction:					
International Building Code	90	60	60	60	60
National Building Code of India	120	90	60	60	-

Both the International Building Code and the South Africa Building Regulations for Fire, SANS 104100-T, provide fire resistance ratings for exterior walls to buildings based on use and occupancy.

Under use and occupancy, the International Building Code divides buildings and structures into groups defining a building's specific use and are numbered based on the perceived risk the building poses to its occupants in the case of a fire (refer Table 4.6).

Table 4.6: Fire resistance in minutes of external walls (IBC, 2015), (SANS 10400-T, 2011)

Occupancy	International Building Code	South African Building Regulations
All occupancies except those mentioned below	-	30
Assembly, Business, Educational, Institutional, Residential, Utility & Miscellaneous	60	60
Factory, Mercantile, Storage	120	60
High hazard	180	120

Studies were conducted by Mena (2012) to determine the char rate for bamboo. He exposed specimens of the species *Guadua angustifolia* to a surface temperature similar to the action of fire using an electric muffle furnace. The test specimens were exposed to a temperature of 425 C for two minutes, and then exposed to a temperature of 524 C for a further two minutes (temperatures of 425 C and 524 C correspond to the temperatures after two and four minutes respectively of the standard time-temperature curve used in full-scale tests as set by the International code ISO 834: *Fire-resistance tests*). Mena reported a char rate of 0.2 mm/minute for *Guadua angustifolia* with a reported density of 700 kg/m^3. This char rate is a low value when compared with timber, which chars at a rate of 0.6 – 0.8 mm/minute (Kaminski, et al., 2016), (SANS 10163-1, 2003). It was also noted that bamboo chars at a fast rate once ignited, although actual figures were not given.

Although bamboo chars at a slower rate than timber, its structural strength will be compromised sooner since it has thin walls and is not a solid element.

For example, using *Guadua angustifolia* as described by Mena, with a char rate of 0.2 mm/minute, let us assume it has a diameter of 120 mm and a wall thickness of 15 mm (values obtained from Chapter 3). Therefore after 35 minutes, the bamboo would have lost 50% of its wall thickness and cross-sectional area, and subsequently a substantial portion of its strength. After 75 minutes, the entire culm wall would have charred.

Although the silica found on the outer skin of the culm walls provides some resistance to fire, once this has been removed due to charring, the fire resistance reduces.

There are several measures available to increase the fire resistance of the bamboo culms. Fire resistance can be addressed during the chemical preservation treatment, by adding fire retardants, such as boron, to the chemical composition. In addition, Janssen (2000) suggested plastering both sides of woven bamboo walls to eliminate the risk of catching fire, while Kaminski (2016) suggested the use of surface treatments or fire-retardant coatings.

The Indian design code (Bureau of Indian Standards, 2015) recommends treating the bamboo culms with a chemical mix of sodium dichromate, zinc chloride, copper sulphate, boric acid, ammonium phosphate and water, as a means of providing protection against fire, while the International Code Council specifies the use of borate solution (ICC Evaluation Service, 2017). Although these chemical treatments might increase the fire resistance of bamboo, there is concern that these substances might release toxic fumes in the process. This area requires further investigation.

Insufficient information was found in the literature to validate the claims that treated bamboo is an acceptable building material regarding fire protection. This area requires in-depth investigation if bamboo is to be considered a viable structural building material.

Although this section only covers the fire resistance of bamboo as a material, attention would need to be given to the fire rating of any materials used in connections, as this would affect the fire rating of the structure.

4.4   Mechanical properties

4.4.1   Axial Compression and Tension

Kaminski (2017) stated that culms in compression fail due to local crushing of the fibres in short elements, bursting of the culm wall in medium length elements and buckling in long elements. Buckling of culms is covered in more detail in Section 4.5.

There are several reports for compressive strength for bamboo culms, but little is said regarding the testing procedures, the actual species involved, or which portions of the culms were used. Pawar (2014) stated that bamboo has a safe working compressive stress of 100 MPa, while Arce-Villalobos (1993) cited various studies on the compressive behaviour of bamboo, with results varying from 27 MPa to 83 MPa. Xiao (2010) reported a compressive strength of 55 MPa, while DeBoer (2000) reported a compressive strength of 64-110 MPa. Some of the works cited reported that compressive strength decreased towards the nodes, which would indicate a possibility of local buckling at those areas.

Naik (2005) used test procedures as per ASTM D 695-96 to determine the compressive strength of bamboo culms. The results varied between 51 MPa for *Bambusa bambos* and 105 MPa for *Melocanna bambusoides*. Failure occurred from micro-buckling of the cellulose fibres leading to kink band formation and shear failure. Naik also reported a compressive strength of 48 - 81 MPa for *Bambusa balcooa*, and 73 - 76 MPa for *Bambusa nutans*.

Similarly, many studies and reports have been done to determine the tensile strength of bamboo, although little description is given regarding the procedures followed, or the species tested.

Arce-Villalobos (1993) cited several studies with tensile strengths between 54 MPa to 342 MPa. Naik (2005) reported strengths between 121 MPa and 210 MPa, depending on species, while Xiao (2010) reported a typical tensile strength of 124 MPa.

A summary of the tensile and compressive strengths of certain species is given in Table 4.7. A complete list is given in Appendix A.

Some of the literature provided single values for the compressive and tensile strengths of the various species, while others provided a range of values. Although it was not stated explicitly, it can be assumed that the single values represent average results, as multiple samples were tested in each instance. Furthermore, where values were given to one decimal place, these could be ascribed to more accurate testing equipment, or more accurate reporting.

Table 4.7: Compression and Tensile strengths of Bamboo species

Bamboo species	Compression strength parallel to the grain (MPa)	Tensile strength parallel to the grain (MPa)	Reference
Bambusa balcooa	48 - 81 (dried)	107 - 228 (dried)	(Naik, 2005)
	51.0 - 57.3 (dried)	n.a.	(INBAR, 2000)
	60.6 (dried)	n.a.	(Bureau of Indian Standards, 2015)
Bambusa nutans	47.9 (dried)	n.a.	(Bureau of Indian Standards, 2015)
	73 - 76 (dried)	115 - 288 (dried)	(Naik, 2005)
Bambusa polymorpha	32.0 - 41.9 (green) 45.2 - 53.4 (dried)	n.a.	(Jansen & Duriyaprapan, 1995)
Bambusa stenostachya	4.1 (dried)	7.6 (dried)	(ICC Evaluation Service, 2017)
	51.0 - 66.3 (dried)	n.a.	(Richard, 2013)
Bambusa vulgaris	25.3	n.a.	(Chaowana, 2013)
	38.6 (green)	n.a.	(Bureau of Indian Standards, 2015)
	41.7 - 48.4	119.6 - 139.1	(Widjaja & Risyad, 1980)
	43.2 - 65.2 (dried)	103.3 - 136.6 (dried)	(Mbuge, 2000)
	na	210.3 - 248.1	(Awalluddin, et al., 2017)
Dendrocalamus asper	31.0	n.a.	(Brink, 2008)
	31.5	n.a.	(Chaowana, 2013)
	na	137.2 - 289.4	(Awalluddin, et al., 2017)
	56.5 - 63.9	204.0 - 221.9	(Widjaja & Risyad, 1980)

Table 4.7 (continued):

Bamboo species	Compression strength parallel to the grain (MPa)	Tensile strength parallel to the grain (MPa)	Reference
Dendrocalamus giganteus	55.6	121.5	(Baba & Bhalla, 2009)
	60.2 - 64.5	183.6 - 196.5	(Widjaja & Risyad, 1980)
	60 - 91 (dried)	160 - 194 (dried)	(Naik, 2005)
Dendrocalamus strictus	42 – 57 (dried)	n.a.	(Jansen & Duriyaprapan, 1995)
	69.1 (dried)	n.a.	(Bureau of Indian Standards, 2015)
	na	106.2 - 185.3	(Ahmad & Kamke, 2005)
	80	160	(Liang & Robinson, 2009)
Gigantochloa apus	n.a.	178.8	(Abdullah, et al., 2017)
	33.9 (dried)	299 (dried)	(PlantUse English, 2016)
Guadua angustifolia	56	140	(Yu, 2007)
Phyllostachys pubescens	56	196	(Yu, 2007)

Note: n.a. = not available

This large variability and lack of clarity regarding tensile and compressive strength values for specific species indicate that this area needs substantial further research. It also indicates the importance of testing and assessing the actual culms to be used in any project, prior to construction.

4.4.2 Column buckling

The shape of the column is important in the stability of the column. A cylindrical shape places the column's outer boundary and its material at a consistent distance from the centre. Hence the column has no side that is weaker and thus more susceptible to buckling than any other side. Loads are equally distributed over the walls of the cylinder.

Richard (2013) performed several experiments on columns to determine their buckling behaviour (refer Figure 4.3); this will be covered in more detail in Chapter 5.

The International Organisation for Standardization (ISO) document ISO 22156: *Bamboo Structural Design* (ISO 22156, 2004), provides guidance for determining the culm buckling load and moment of inertia. ISO 22156, the National Building Code of India (2015) and the ICC (2017) all specify the use of the Euler buckling equation but reducing the moment of inertia by 10% to account for variation in diameter along the length of the culm, i.e. the effect of the taper. Additionally, these three references state that column designs should consider bending stresses due to induced deflection, eccentricities and initial curvature.

Figure 4.3: Column buckling test result (Richard, 2013)

4.4.3   Flexure

In the available literature, bending or flexural strength is measured in terms of the Modulus of Rupture (MOR), which is a measure of a material's strength before rupture or failure in bending.

Similar to the previous mechanical properties, the values given for bending strength vary widely. The results are summarised in Table 4.8 and discussed briefly below.

Table 4.8: Flexural strength of bamboo species

Bamboo species	Modulus of Rupture / Flexural strength (MPa)	Reference
Bambusa balcooa	65.4 (green)	(Bureau of Indian Standards, 2015)
	69.6 - 92.6 (dried)	(INBAR, 2000)
	80.3	(Anokye, et al., 2016)
	116 - 173 (dried)	(Naik, 2005)
Bambusa nutans	52.4 (dried)	(Bureau of Indian Standards, 2015)
	84	(Janssen, 1981)
	87.7	(Anokye, et al., 2016)
	161 - 236 (dried)	(Naik, 2005)

Table 4.8 (continued):

Bamboo species	Modulus of Rupture / Flexural strength (MPa)	Reference
Bambusa polymorpha	37.3 - 46.9 (green) 41.4 - 56.6 (dried)	(Jansen & Duriyaprapan, 1995)
Bambusa stenostachya	2.5 - 8.0	(Sharma, 2010)
	10.3 (dried)	(ICC Evaluation Service, 2017)
Bambusa vulgaris	41.5 (green)	(Bureau of Indian Standards, 2015)
	62.3	(Chaowana, 2013)
	86.0	(Anokye, et al., 2016)
	92.1 - 158.6 (dried)	(Mbuge, 2000)
	107.5 - 128.6	(Widjaja & Risyad, 1980)
Dendrocalamus asper	85.7	(Chaowana, 2013)
	103	(Brink, 2008)
	105	(Anokye, et al., 2016)
Dendrocalamus asper	157.8 - 174.2	(Widjaja & Risyad, 1980)
Dendrocalamus giganteus	175.8 - 288.0	(Widjaja & Risyad, 1980)
	173 - 224 (dried)	(Naik, 2005)
Dendrocalamus strictus	50 – 153 (dried)	(Jansen & Duriyaprapan, 1995)
	118.4	(Anokye, et al., 2016)
	119.1 (dried)	(Bureau of Indian Standards, 2015)
Gigantochloa apus	74.9 (dried)	(PlantUse English, 2016)
	36.9 - 94.7 (dried)	(Nugroho, et al., 2018)
Guadua angustifolia	72.6 (dried)	(Gnanaharan, 1994)
	74.0	(Yu, 2007)
Phyllostachys pubescens	129.1	(Yu, 2007)

Janssen (1981) stated that the bending strength of dry bamboo is approximately 1.5 times the strength for green bamboo. However, no specific values were given.

Naik (2005) reported bending strength for bamboo varying from a minimum 49 MPa for *Dendrocalamus hamiltonii* to a maximum of 236 MPa for *Bambusa nutans*.

Gnanaharan (1994) tested straight culms of the species *Guadua angustifolia*, using different types of test specimens. Four-point bending tests were applied to long specimens with a span of 3000 mm, while three-point bending tests were applied to short specimens with a span of 700 mm. The results of the long specimens gave an average Modulus of Rupture of 72.6 MPa. The results from the short specimens were substantially lower (average MOR = 56.2 MPa) and varied depending on the location of the specimen in the culm. This variation in results was attributed to the fact that smaller specimens are liable to get crushed at the point of load transfer at lower loads, resulting in lower ultimate strengths. Gnanaharan concluded that results obtained from bending tests with short spans did not reflect the true potential of bamboo.

Four-point bending tests for full-culm bamboo are given in the ISO test method document (ISO 22157-1, 2004), where failure is considered to be flexural and the "modulus of rupture" or "flexural modulus" is calculated. Gottron (2014) disputes this failure method, stating that most culms tested in flexure display a splitting mode of failure as a result of longitudinal shear failure. He further states that this is the critical failure mode associated with bamboo structures and connections (refer Figure 4.4 and Section 4.5).

Figure 4.4: Typical splitting failure in bamboo culms (Janssen, 2000)

4.4.4 Shear

In many instances, the limiting factor for shear strength is the strength of the lignin matrix. When compared with timber, the hollow cross-section of the culm has less area to resist shear forces than a solid piece of timber of the same diameter (Richard, 2013).

The available information regarding the shear strength of bamboo varies from 1.3 MPa for *Bambusa stenostachya* (ICC Evaluation Service, 2017) to 15 MPa for *Dendrocalamus strictus* (Jansen & Duriyaprapan, 1995).

Shear strength of bamboo culms is commonly tested using the 'bow-tie' test method, as presented by ISO 22157-1 and 22157-2 (ISO 22157-1, 2004) (refer Figure 4.5) The test places a segment of the culm into longitudinal compression between two plates arranged like a bow-tie. The plates are placed at 90 degrees relative to each other at either end of the specimens, creating four longitudinal shear planes along the length of the specimen along which failure can occur (refer Figure 4.6). However, it was not clear whether this method of testing was used in the literature reporting on the shear strength of bamboo.

Naik (2005) recorded a transverse shear strength of 10.1 - 13.1 MPa for *Bambusa balcooa*, and 8.3 - 14.3 MPa for *Bambusa nutans*, while the ICC (2017) calls for an allowable design shear stress of 1.3 MPa for *Bambusa stenostachya* culms used for structural purposes.

Jiang (2012) tested *Phyllostachys pubescens* (Moso) bamboo samples for sensitivity to moisture content. He concluded that shear strength parallel to the grain was the most sensitive to moisture content, varying in strength by 3.1%. He ascribed this sensitivity to the weakening of the bond between the parenchyma cells and the longitudinal fibres.

Figure 4.5: Bow-tie shaped plates used in shear tests for bamboo (Trujillo & Lopez, 2016)

Figure 4.6: Loading device for shear testing of culm specimens (ISO 22157-2, 2004)

Some typical values for shear strength are given in Table 4.9.

Table 4.9: Shear strength of Bamboo species

Bamboo species	Shear strength (MPa)	Reference
Bambusa balcooa	10.1 - 13.1 (dried)	(Naik, 2005)
Bambusa nutans	8.3 - 14.3 (dried)	(Naik, 2005)
Bambusa polymorpha	n.a.	-
Bambusa stenostachya	1.3 (dried)	(ICC Evaluation Service, 2017)
Bambusa stenostachya	3.3	(Gottron, 2014)
	5.0 - 10.7 (dried)	(Richard, 2013)
Bambusa vulgaris	4.0	(Chaowana, 2013)
Dendrocalamus asper	5.4	(Chaowana, 2013)
	7	(Brink, 2008)
Dendrocalamus giganteus	8.7 - 12.0 (dried)	(Naik, 2005)
Dendrocalamus strictus	13.2	(Liang & Robinson, 2009)
	13 – 15 (dried)	(Jansen & Duriyaprapan, 1995)
Gigantochloa apus	7.7 (dried)	(PlantUse English, 2016)
Guadua angustifolia	9	(Yu, 2007)
Phyllostachys pubescens	9.7	(Gottron, 2014)
	13.9	(Yu, 2007)
	14.2 (dried)	(Richard, 2013)

Note: n.a. = not available

Shear failure is critical when considering joints to connect bamboo culms. In most methods, holes are made in the culm wall, with the fastener being placed through this hole. Thus, when in use, the forces on the fasteners will be transferred to the culm walls in shear (refer Figure 4.7 and Figure 4.8). Bamboo connections will be dealt with in more detail in Chapter 5.

Figure 4.7: Failure in shear in the direction of fibres (Tönges, 2002)

Figure 4.8: Failure of bolted connections (Sharma, 2010)

4.5   Failure modes

Columns typically fail by crushing or by buckling, such as when the load applied to the longitudinal axis of the column exceeds the compressive strength of the material of the column (crushing) or when the load leads to deformation of the column, leading to instability (buckling).

An important aspect to consider for bamboo construction is that failure in bending does not necessarily indicate structural failure. The Technical University of Eindhoven conducted long-term continuous bending tests on bamboo from 1981 to 1988, and they concluded that when bamboo fails in bending, the longitudinal fibres remain intact, unlike timber specimens. However, the lateral bond between the fibres fails, and the circular form of the cross-section is no longer effective at providing resistance (Janssen, 2000) (refer Figure 4.9) The nodes and diaphragms prevent the spread of the longitudinal cracks over the entire length of the culm. This failure behaviour is important as "failure" does not equal immediate collapse of the structure, thus providing a measure of safety or time to escape the structure before entire collapse occurs. This failure mode is corroborated by Sharma (2010).

Figure 4.9: Typical failure in bending in bamboo culms (Janssen, 2000)

Gottron (2014) describes various modes of failure of a bamboo culm. The first failure type (Mode I) is as a result of transverse opening tension (refer Figure 4.10a). The resistance of bamboo to this type of failure is dependent on its tensile strength perpendicular to the longitudinal fibres.

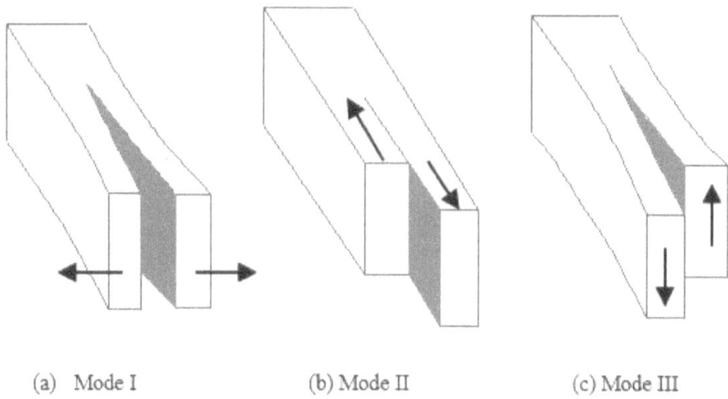

Figure 4.10: Three modes of failure (Gottron, 2014)

The second type of failure (Mode II) is the result of in-plane shear along the length of the culm (refer Figure 4.10b). According to Gottron, this type of failure does not typically determine failure in bamboo culms. Instead, failure tends to be a combination of the two above-mentioned types (refer Figure 4.10c), referred to by Gottron as "mixed-mode failure" (Mode III), where shear occurs perpendicular to the fibres of the culm (Gottron, 2014).

4.6   Durability

Bamboo has less natural durability than most timber. The hollowness of the bamboo culm offers a safe hiding place for insects. In addition, if insects or fungi attack the outer layer of timber, to a depth of 3 mm, the cross-section of the timber will still be in good condition. However, if the outer wall of bamboo is reduced by 3 mm, it could lose approximately a third of its thickness. The high inherent moisture content makes bamboo susceptible to fungal attack (Janssen, 2000).

Furthermore, bamboo is susceptible to moisture, both from the environment and from moisture present in the soil. Architects such as Simon Velez recommend a 'good shoes, good hat' approach, where the bamboo structures do not penetrate the ground, and the structures are provided with a suitable roof. This is covered in more detail in Chapter 5.

A rough guideline on the service life of untreated bamboo was presented by Janssen (2000), as follows:

- 1-3 years exposed to the environment and in contact with soil

- 4-6 years under cover and free from soil contact
- 10-15 years under good storage and use conditions (Janssen, 2000)

With these short service lives, bamboo would only be suitable for temporary structures, as a typical design life, and thus expected service life, for most buildings is 50 years. Preservation of the culms can increase these life spans considerably to 25 years or more.

Preservation or treatment of the culms can shift the use of bamboo in construction from the realm of traditional construction and the perception of "poor man's timber", to modern engineered construction, with engineered connections. Furthermore, certain design codes recommend some form of preservative treatment (ISO 22156, 2004), (Bureau of Indian Standards, 2015).

However, bamboo is not naturally conducive to preservation. The outer skin has a silica layer which protects the bamboo from insects and rain but also prevents the ingress of preservatives. The inside similarly has an impermeable waxy layer. Thus, a preservative can only enter the culm through the exposed ends of the conducting vessels in the culm wall, which make up 10% of the culm wall. In addition, these vessels close within 24 hours after harvesting, leaving a small timeframe in which to affect preservation.

Preservation of bamboo is not necessarily toxic to humans, as the aim is merely to stop borers and termites from destroying the bamboo culms. There are two main types of preservation – traditional and chemical. Traditional preservation consists of curing, smoking, soaking and lime-washing. Chemical treatment involves the use of chemicals containing the element boron, such as boric acid, or copper-chrome-boron (CCB). Chemical treatments used in the past, such as pentachlorophenol (PCP) or chromated copper arsenate (CCA), have been banned in most countries, as they contain arsenic and chromium, both of which are known to be carcinogenic (Kaur, et al., 2016).

The choice of preservation method depends on the size and scale of the project, as well as the required level of durability. Other factors include budget and availability of materials and supplies.

4.6.1 Traditional preservation methods

In smoking or curing, the bamboo is placed above a fireplace so that the smoke and heat dry out the culm. The smoke makes the rind unpalatable to insects, thus preventing attack. A variation on this method is the heating of the culms in kilns at temperatures of 150 °C. The heating changes the structure of the outer zone of the culm, making it more resistant to insect attack (Bamboo Village Hawaii, 2004).

Bamboo culms can also be immersed or soaked in water for four to twelve weeks, which causes the nutrients to be removed (refer Figure 4.11). Saltwater is not considered suitable for this, as the salt stays in the culms, which can encourage the retention of moisture and the growth of fungi in the culms. The culms can also be cooked or boiled, which reduces the starch and nutrients in the culms. Both methods reduce the nutrients which are attractive to insects and beetles and thus reduce the likelihood of insect attack. Due to the anatomy and structure of the bamboo culm, there is little radial movement of liquids through the culm. This lack of movement creates problems for preservation techniques using liquids, as the penetration pathways are through the cut ends of the culms. For this reason, it is not necessary to submerge the bamboo, only the cut ends.

Figure 4.11: Leaching of nutrients from bamboo culms (Schröder, 2012)

Bamboo can also be painted with slaked lime to enhance the final appearance and prevent moisture penetration. Bamboo structures can also be tarred and plastered before whitewashing.

4.6.2   Chemical preservation methods

Chemical treatment often consists of soaking in a boron salt solution. This treatment method is useful in protection against fungi, particularly in wetter climates. The boron acts as a feeding toxin to the insects. It changes the pH level of the intestines such that the intestinal bacteria can no longer function, and the insects die. The ICC (2017) states that culms used for structural purposes should be treated with a borate solution.

Although the culms can also be treated with insecticide, this method is not recommended as the insecticides often present a danger to the environment. Phosphate-based insecticides have been reported as having allergenic potential, while others have been branded as carcinogenic.

A popular method of preservation is the Pressurised Sap Displacement method, also known as the Modified Boucherie method (refer Figure 4.12). This method consists of pumping a chemical liquid into the open vessels in the walls of the culms, replacing the fresh sap. This action needs to occur within 24 hours of harvesting before the conducting vessels close. The advantage of this method is that the liquid used in the process can be recycled (Janssen, 2000).

Figure 4.12: Boucherie method of chemical treatment (Janssen, 2000)

An alternative chemical treatment is known as dip diffusion. The culm is immersed or dipped in the preservative so that a slow diffusion takes place. Janssen (2000) notes that this method can only be applied to split or sawn bamboo strips, since whole culms will not allow the preservative to penetrate (Janssen, 2000).

4.7   Bamboo Building Products

In traditional bamboo construction, whole culms constitute the poles, which in turn form structural frames which are able to withstand compression and bending in their members. One of the problems encountered with using full culms is their natural variation in geometry, which prevents standardisation of the connections as well as variation in their mechanical properties. This lack of standardisation in turn leads to problems with quality control of the construction. A solution to this variable is the production of engineered bamboo products, where raw bamboo is processed to form composite materials.

There are numerous structural bamboo products in the current market. Two examples are Lamboo WeaveCore, an engineered bamboo plywood developed for the University of Illinois 2009 Decathlon Project (Aijazi, 2013), and GluBam, a composite of bamboo strips used in a prototype house in China in 2008 (see Figure 4.13).

All parts of the bamboo culm can be processed to form a multitude of different products for various uses, including structural applications, as illustrated in Figure 4.14.

Figure 4.13: GluBam concept house in Changsha, Hunan Province, China, 2009 by researchers at the University of Southern California (Aijazi, 2013)

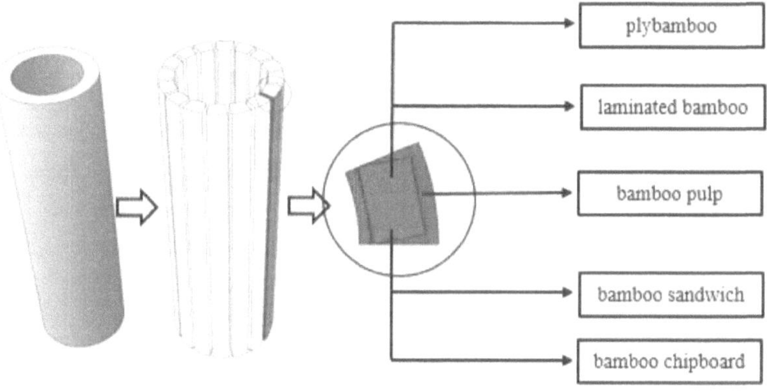

Figure 4.14: Industrial bamboo products (Yu, 2007)

A detailed review of bamboo products is outside the scope of this book.

4.8 Bamboo as a sustainable construction technology globally and locally

With the current focus on the environment and sustainable construction, many architects include natural materials in their buildings, particularly in the finishes. In 2018, the Beaufort West Hillside clinic was completed (refer Figure 4.15), incorporating several sustainable design features, such as rammed earth walls, underground thermal rock stores, and a profusion of timber structural members (WCGTPW, 2018).

Figure 4.15: Beaufort West Hillside Clinic during construction (WCGTPW, 2018)

As in the Beaufort West Clinic, timber is often used as a structural member in the place of steel, as well as in the finishes. Although much has been done regarding sustainable forestry practices to ensure the sustainable use of timber as a construction material, other natural resources are also being investigated, such as bamboo.

Bamboo plants meet many of the criteria for sustainable, renewable construction materials. They have a fast growth rate; require minimal processing and many suitable varieties are available for construction (van der Lugt, et al., 2009). They reach maturity in 3 to 6 years, from when they become suitable for use as structural members in construction. In comparison, timber used for construction purposes only reaches maturity in 15 – 20 years, in some cases even longer, depending on the timber species.

4.8.1 Environmental Assessment of Bamboo construction

Although many publications state that bamboo is a sustainable building material, comparing favourably with other building materials, the quantitative proof appears to be lacking. In 2003, a study was published on the environmental performance of bamboo when used in a specific application (van der Lugt, et al., 2003).

Van der Lugt (2003) compared the environmental impacts of a bamboo pedestrian bridge with the conventional materials used for such a structure by means of a cradle-to-grave Life Cycle Assessment (LCA). The bamboo bridge in the Amsterdam Woods in the Netherlands (refer Figure 4.16) was made from the species *Guadua angustifolia* grown in Costa Rica in Central America and imported into the Netherlands. For each element of the bridge, the bamboo culms were compared with building materials typically used for each application, namely steel, timber (*azobe* and *robinia*) and concrete.

Figure 4.16: Bamboo pedestrian bridge in the Amsterdam Bos (van der Lugt, et al., 2005)

Van der Lugt found that the environmental cost of transporting the bamboo culms to the Netherlands was approximately 93% of the bamboo's total environmental cost.

To compare the different building materials, van der Lugt divided the environmental cost of each material with its expected lifespan, to obtain the relative annual environmental costs. He represented this as an index, where the environmental costs of an alternative were divided by the environmental cost of the alternative with the lowest environmental cost, and the result multiplied by 100 (refer Figure 4.17).

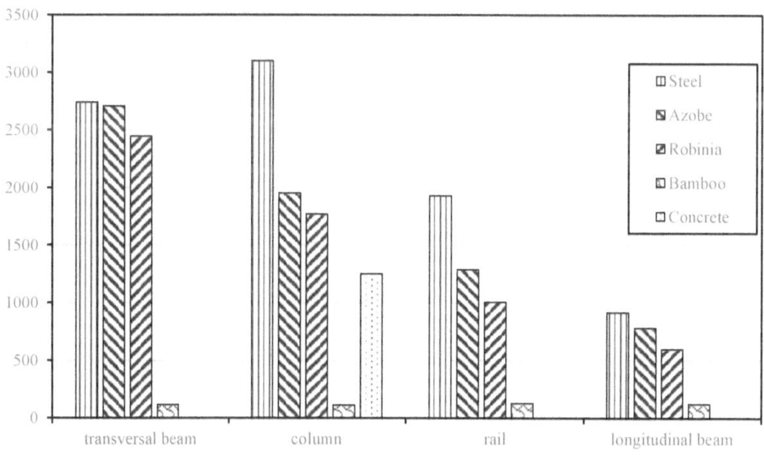

Figure 4.17: Index of annual environmental costs of various materials (van der Lugt, et al., 2003)

He concluded that in these applications, bamboo could be considered the most sustainable alternative for all elements, despite the additional environmental costs caused by the transportation factor. He ascribed this to bamboo's low mass to strength ratio and the simple production process for producing the culms. He further noted that the bamboo option compared less favourably in the longitudinal beam member as opposed to the transverse beam member, since the longitudinal beam required four culms whereas the transverse beam only required two.

Van der Lugt made a similar comparison between the building materials, based on purchasing and production costs. The cost of each material was then divided by its lifespan, and he concluded that steel was a more cost-effective solution than bamboo, as a result of bamboo's higher labour costs and shorter life span (refer Figure 4.18) (van der Lugt, et al., 2003).

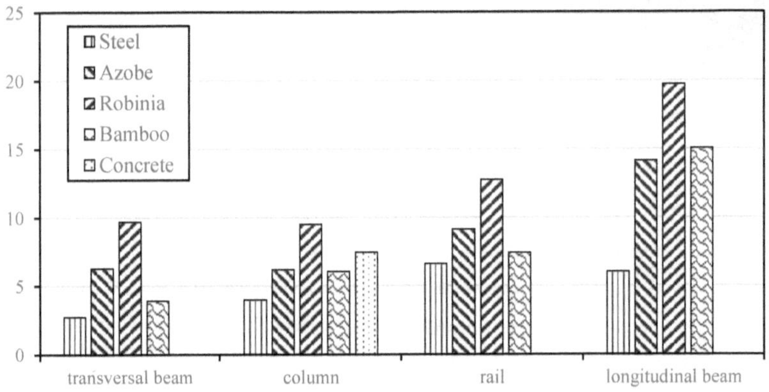

Figure 4.18: Annual costs of various bridge elements and materials (van der Lugt, et al., 2003)

Murphy (2004) conducted a study to determine the environmental impact of the structural elements of a house built from bamboo. He used the Life Cycle Analysis (LCA) method on a 35m² house built from the bamboo species, *Guadua angustifolia*. The house was built using an engineered bahareque construction method, which consists of a plastered bamboo frame on reinforced concrete foundations, similar to the traditional wattle-and-daub type housing.

The results of the LCA of the bamboo house were compared with that of a similar house built using conventional masonry construction. The basic building parameters are given in Table 4.10 (Murphy, et al., 2004).

The elements of the two buildings were assessed utilising means of Eco-indicators (EI99) in order to compare all elements on the same basis. Eco-Indicator 99 (EI99) is a life cycle impact assessment tool developed by PRé Consultants where scores are assigned to each material or process used. The scores obtained can thus be compared on an equal basis to determine which material or product has a greater or lesser environmental impact.

Table 4.10: Basic building parameters of *Guadua* (actual) and masonry (theoretical) houses (adapted from Murphy, et al., 2004)

Parameter	*Guadua* house	Masonry house
Floor size	35 m²	35 m²
No of floors	2	2
Wall structure	*Guadua* / Timber / Cement	Brick & concrete
First-floor structure	Timber	Concrete
Roof structure	Light fibre cement	Light fibre cement
Basic fittings (windows, doors) and services (water, electrics) considered equivalent and thus ***excluded*** from the analysis		

The nine environmental impact categories used by Murphy are indicated in Figure 4.19. The results indicated that the bahareque house had, on average, an environmental impact approximately 40% lower than that of the conventional masonry house (Murphy, et al., 2004) (refer Figure 4.19). An explanation of typical LCA methodologies is given in Chapter 2.

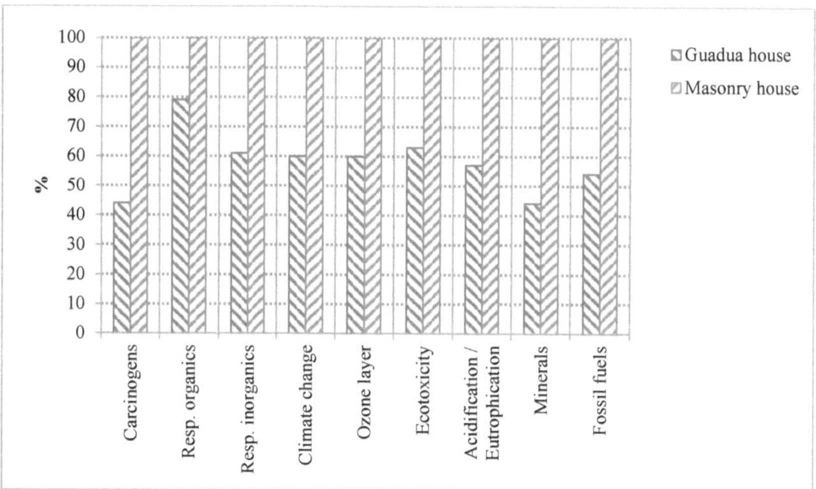

Figure 4.19: Relative LCA environmental impact of the *Guadua* and masonry houses (Murphy, et al., 2004)

Murphy further elaborated that 95% of the environmental impact for the *Guadua* house came from the foundations and the wall plaster, both of which contained cement. He did a further analysis, where the foundation and roof systems were excluded, as these were common to both house types. Thus, only the walls, floors and stairs were compared, with the results given in Figure 4.20. This further analysis indicated that the bamboo wall system had less than 50% of the environmental impact than that of the masonry system.

Murphy concluded that the structural elements of the bamboo house had approximately half the environmental impact of a conventional masonry wall system, due in part to the light weight of the material, its simple processing requirements and the short transport distances required in this instance.

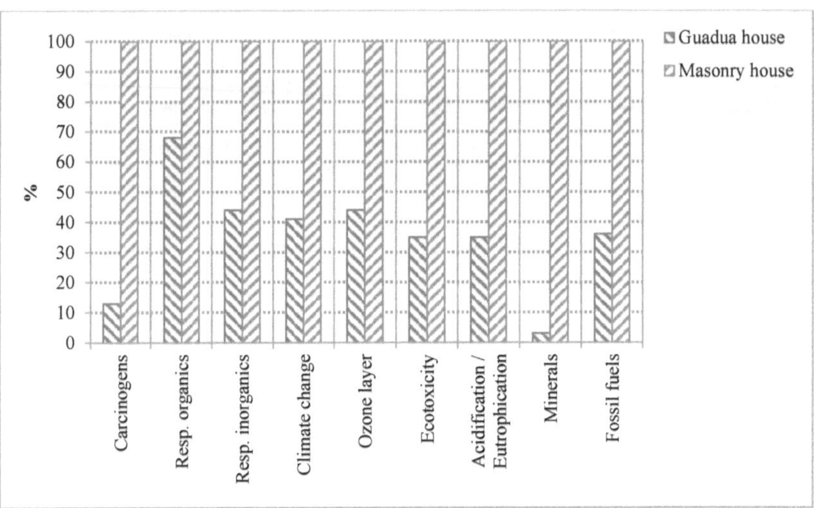

Figure 4.20: Relative LCA environmental impact of the wall/floor/stair elements of the *Guadua* and masonry houses (Murphy, et al., 2004)

In 2011, a study was undertaken by undergraduate students at the University of Pittsburgh, comparing the LCA of bamboo and timber portal frames. Bamboo produced locally and domestically, i.e. not in a plantation, was deemed to be a more sustainable option than timber, based on the functional unit of equivalent lateral stiffness (Richard, 2013).

A second study was done in 2012 by undergraduate students at the University of Pittsburgh, comparing LCA of bamboo and glass fibre reinforced polymer (GFRP) gridshell structures used for rapidly deployable relief shelters in response to natural disasters (Richard, 2013). The results indicated that if bamboo is located in a favourable climate, grown and harvested in a sustainable manner, with a limited delivery transport distance, bamboo can be a suitable sustainable material alternative to GFRP for the gridshell structures considered. These results could be extrapolated to apply to all bamboo construction, i.e. bamboo can be considered a favourable sustainable material alternative if grown and harvested sustainably and within a certain radius of the construction work, thus keeping transport to a minimum.

In 2013, Richard compared the environmental impacts of a bamboo column with the impacts of timber and steel column alternatives with a similar capacity. Unlike the studies done by van der Lugt et al. (2005, 2009), no weighting was given to the impact categories, with the result that no clear sustainable option was found between the three materials (Richard, 2013). However, the study did show that locally grown bamboo has a better environmental performance, due to the reduced transport necessary. This indicates that the weighting applied in comparisons could influence the results and should be considered carefully.

Stematis and Gangyi (2019) compared a steel frame structure and a bamboo frame structure for a tourist building in Zhengjia Shan, in the Hubei province of China. They concluded that the use of steel had a negative environmental impact, as it emitted approximately 6400 kg of CO_2. In contrast, the bamboo version would have stored approximately 14300 kg of CO_2 over the building's lifetime. Although the total length of bamboo poles that would have been required was 1750 m, which was double the steel requirement of 875 m, the transport of the bamboo poles would have been significantly less. The bamboo was available locally and would have entailed transport of 2 km, whereas the steel would have been transported over 60 km. Thus, although the total steel mass was less, transporting 1488 kg of steel over 60 km had a more significant environmental impact than transporting 1925 kg of bamboo over 2 km.

Stematis and Gangyi further noted that constructing the building in steel took six days, whereas construction in bamboo would have taken approximately 30 days. This additional time and labour cost are not reflected in the environmental assessment but could have a significant effect on the choice of materials.

They quoted work done by Trujillo and Malkowska (2017) stating that a steel tube as used for the tourist building would emit 3.33 kg of CO_2 to the environment per meter length, while bamboo poles would only emit 0.33 kg of CO_2 per meter length. Furthermore, the steel does not store any carbon while the bamboo poles store 10 kg of CO_2 per kilogram of mass. Thus, the use of steel in the building would have emitted approximately 6400 kg of CO_2 into the environment. In contrast, the bamboo version would have stored approximately 14300 kg of CO_2 over the building's lifetime (Stamatis & Gangyi, 2019).

4.8.2 Environmental Assessment of Bamboo products

Although a detailed review of bamboo products is outside the scope of this book, a brief review of their environmental impact is included for comparison with bamboo culms.

Bamboo is a versatile material with a variety of uses in construction. However, as a natural material, it has variations in geometry and quality, which make standardisation of structural design difficult. To combat this uncertainty and variability, a variety of bamboo boards and panels have been developed, with varying strengths and applications (refer Figure 4.21).

The advantage of bamboo products is that material defects or variations in the original culms can be removed, and greater quality control can be exercised with the engineered products, leading to greater consistency in their structural properties.

Figure 4.21: Plybamboo boards (van der Lugt & Vogtländer, 2015)

However, it could be assumed that the increased processing required to produce these boards and panels also increases the environmental footprint and thus the ecological cost.

Van der Lugt and Vogtländer (2015) used an LCA approach to compare the environmental impact of industrial bamboo products found in Western Europe with more commonly used materials such as European softwood and concrete. The study used information obtained from MOSO International BV in the Netherlands and stated that the results are not necessarily typical for other industrial manufacturers of bamboo products.

The bamboo species used in the study was *Phyllostachys pubescens*, also known as Moso bamboo, with a density of 700 kg/m^3, a diameter of 100-120 mm and a wall thickness of 9 mm. It was assumed that at the end of the useful life of each product, 90% of the bamboo products would be incinerated in an electrical power plant, and 10% would end up in landfill. In addition, the carbon sequestration of the various products was considered. The LCA was based on a cradle-to-gate plus end-of-life analysis, where the usage phase was excluded because emissions in that phase were considered negligible (refer Figure 4.22).

The results of this study are given in Figure 4.23, indicating that most bamboo products have a negative carbon footprint over their life cycle, if production parameters are optimised. Thus, the credits gained through carbon sequestration and from burning the products to produce electricity at the end of the life cycle outweighed the emissions caused by the transport and production processes.

Vogtländer concluded that the two highest contributors to the carbon footprint of the bamboo products were energy consumption and transport. He suggested that using local bamboo species instead of importing the bamboo, such as was used in the study, would lower the carbon footprint of such products.

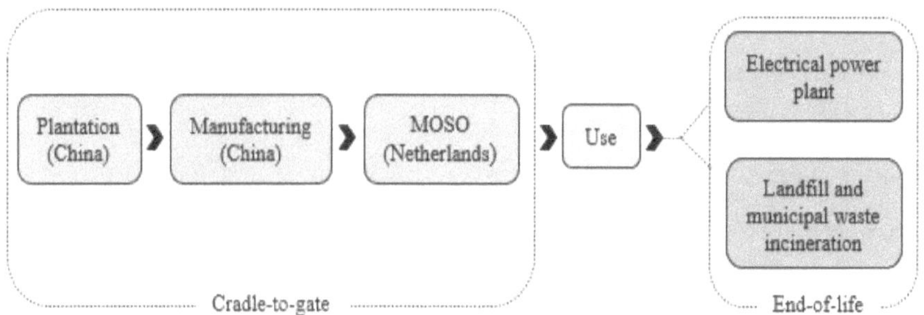

Figure 4.22: LCA system boundaries assumed (van der Lugt & Vogtländer, 2015)

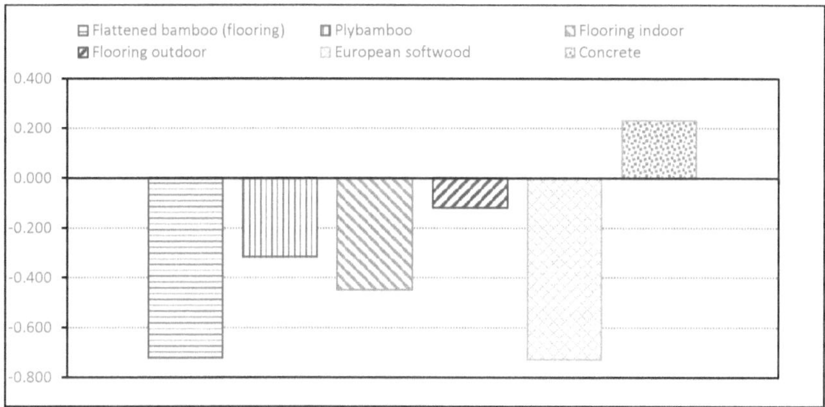

Figure 4.23: Carbon footprint over life cycle (CO$_2$-eq/kg product) (extracted from data) (van der Lugt & Vogtlander, 2015)

From the results given, the European softwood has less environmental impact than the bamboo products. However, although information was not given in the study, it can be assumed that the transport of the European softwood was minimal, while the bamboo for the bamboo products had to be transported from China. As stated by Vogtländer (2014), if a local bamboo species had been used with minimum transport, the carbon footprint of the bamboo products would have been much lower.

Although details were not given of the equivalent carbon footprint of bamboo culms of the same species, it can be deduced that the carbon footprint of the culms would be lower than the bamboo products, due to the minimal processing required. Furthermore, if a local species of bamboo were used, thus reducing the environmental impact of the transportation involved, the carbon footprint of the products and by deduction, the culms, would be lower than the local wood species included in this study.

Although this study was for a specific species of bamboo and bamboo products from one specific company, it can be deduced that bamboo and bamboo products could have a lesser carbon footprint than local wood or other building materials. However, this would need to be verified in each particular case, using the appropriate input data.

4.8.3   Embodied Energy of Bamboo

Manandhar et al. (2019) reported that bamboo had the lowest energy requirement for production, when compared with steel, concrete and timber (refer Table 4.11). Janssen (1981), who originally published these figures, stated that the figures given were not exact, but were meant to indicate an order of magnitude. Furthermore, the embodied energy figure for bamboo was presented with a question mark, indicating that the published figure was based on certain assumptions such as processing and preservation methods, and would not necessarily be applicable in all instances. Therefore, these figures should be regarded as indicators and not absolute values.

Table 4.11: Energy requirements for the production of construction materials (Manandhar, et al., 2019), (Janssen, 1981), (Kibert, 2016)

Building Material	Density (kg/m³)	Embodied energy (MJ/kg)
Concrete (0% fly ash)	2400	0.8 (Janssen, 1981) 0.55 (Kibert, 2016)
Steel	7800	30 (Janssen, 1981) 35.4 (Kibert, 2016)
Wood	600	1 (Janssen, 1981) 10 (Kibert, 2016)
Bamboo	600	0.5? (Janssen, 1981)

4.8.4   Carbon sequestration of Bamboo

Carbon sequestration is the storage of carbon dioxide, either by natural or by engineering means (Kibert, 2016). Biogenic carbon dioxide (CO_2) is the carbon dioxide that is captured in the biomass of a plant during its growth. Thus, as a plant, bamboo stores carbon dioxide while it is growing. Once it is harvested and used in construction, this carbon is stored in the bamboo until the end of the building or structure life, when the bamboo degrades or is burned. Carbon sequestration can be included in an LCA if the bamboo is burned to generate heat or electricity at the end of its life (van der Lugt & Vogtländer, 2015).

Due to the high growth rate of bamboo, Van der Lugt et al. (2009) reported that bamboo plantations are proficient in the sequestration of carbon dioxide, with the carbon constituting approximately half the biomass (dry weight) of the raw bamboo material. The bamboo plant has both green leaves and a green stem, thus having a greater surface area available for photosynthesis. As the bamboo plant grows, it converts carbon dioxide from the air to plant carbohydrates through the process of photosynthesis and emits oxygen into the air (refer Figure 4.24). Bamboo is made up of cellulose and lignin, both of which contain carbon. Thus, bamboo needs to absorb large amounts of carbon to grow. It has been reported that the Guadua bamboo plantations in Costa Rica absorb 17 tons of carbon per hectare per year (Janssen, 2000). Although the amount of carbon dioxide sequestered by bamboo is small in comparison to the amount generated each year, its contribution to improving the environment should not be discounted.

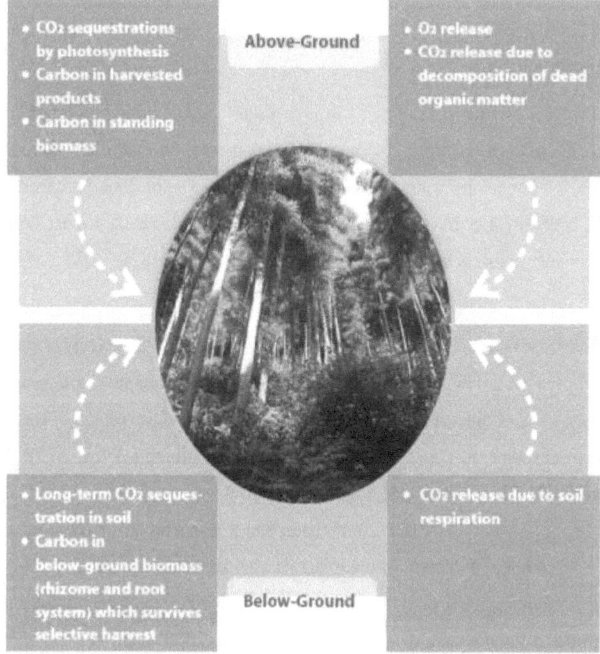

Figure 4.24: The Bamboo Carbon Cycle (Kuehl & Yiping, 2012)

Nath et al. (2015) reported on several studies conducted between 1978 and 2014 by various authors on carbon storage and sequestration by woody bamboo species. The average mean carbon storage was 30 - 121 Mg/ha, while the sequestration rate was reported as 6 - 13 Mg/ha/year (Nath, et al., 2015). In contrast, Huy and Long (2019) reported carbon storage of 94 - 392 Mg/ha, which is substantially higher (Huy & Long, 2019). Nath's findings indicated that certain species had higher carbon storage than others, such as *Guadua angustifolia* in Bolivia with a carbon storage of 100 Mg/ha and *Bambusa bambos* in India with 121 Mg/ha. It would thus appear that the carbon storage, and supposedly the carbon sequestration as well, varies per species of bamboo. Furthermore, as local climate and environmental factors are known to affect the growth rate and properties within a species, it could be presumed that this would affect the carbon storage as well. Thus, the carbon storage capacity might vary within a species, as well as varying between species.

4.8.5   Costs of Bamboo Construction

Bamboo is often seen as a cheap alternative construction material when compared with conventional materials such as concrete or steel. Although this perception may be valid where bamboo is locally available, in areas where bamboo is not grown, the transportation costs to get the bamboo to site could significantly increase the overall cost. This cost is not only monetary, but is considered an environmental cost, as can be seen where transport is included in the LCA studies. These costs would apply to all building materials, not solely to bamboo.

In 1986, the National Bamboo Project in Costa Rica started a low-cost self-build construction program, using the local bamboo species, *Guadua angustifolia*, as the primary construction material. Quintans (1998) reported that it cost USD 83/m² to build these bamboo houses, reportedly 20% less than a conventional house. (Quintans, 1998) In that location, a concrete block wall would have cost USD 38/linear meter while a bamboo panel constructed from local bamboo species cost USD 26.25/linear meter.

A 3-unit classroom block was built from bamboo in Fumesua, near Kumasi, Ghana, in 2003 (refer Figure 4.25). The cost at that time was equivalent to USD 9800 (R 167 000 in 2003), whereas a similar structure built with conventional materials in the same area would have cost the equivalent of USD 16250 (R 277 000 in 2003). This cost represented an approximate saving of 40% (Solomon-Ayeh, 2010). It was noted that the cost of materials for the bamboo building was approximately 70% of the total cost, with the remainder of the cost being ascribed to labour, supervision and transportation. It was further noted that the cost of labour as a percentage of the cost of materials was higher for the bamboo building, but no figures were provided. For countries with a higher labour cost, this would negatively affect the overall cost of such construction and would need to be taken into consideration when costing such buildings.

Figure 4.25: Bamboo school in Fumesua, Ghana (Solomon-Ayeh, 2010)

Also, for this construction, a local bamboo species was used, minimising the transport costs of the material. For areas where a suitable bamboo species is not available locally, the transport costs of such would increase the overall material costs on such projects.

Minimal and simple tools are required for construction, similar to typical carpentry tools (United Nations Industrial Development Organization (UNIDO), 2008) (City University London, 2005), with no special equipment required (refer Figure 7.14). As well as reducing machinery costs, this would reduce the construction time required, and hence costs.

However, the joining skills for bamboo are different to timber, thus requiring training before construction begins (Milani, 2005), which would increase the cost of a project, although this knowledge and skill could then be transferred to future projects.

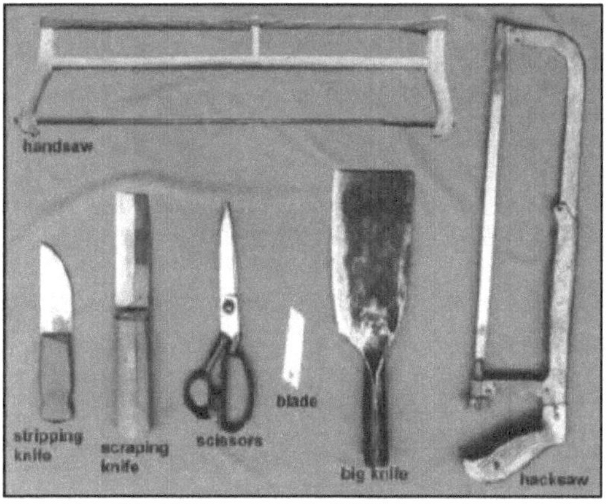

Figure 4.26: Hand Tools for Bamboo Construction (United Nations Industrial Development Organization (UNIDO), 2008)

4.9   Conclusions and recommendations

Numerous tests have been done by various researchers on many species of bamboo to determine their material and structural properties, with large variations in their results. This does not discount the results, but rather indicates the unique nature of bamboo material. Each culm grows differently depending on the amount of water, sun and nutrients it receives, resulting in unique culm properties within a species. In addition, these tests have been done in different places, with differing test conditions, resulting in different results.

The geometrical peculiarities and unique material properties of bamboo affect the design of bamboo structures and building. This will be covered in more detail in Chapter 5.

4.10 References

Abdullah, A. H. D. et al., 2017. Physical and mechanical properties of five Indonesian bamboos. *IOP Conference Series: Earth and Environmental Sciences,* Volume 60.

Ahmad, M. & Kamke, F., 2005. Analysis of Calcutta Bamboo for Structural Composite Materials: Physical and Mechanical Properties. *Wood Science Technology,* September, 39(6), pp. 448-459.

Aijazi, A. N., 2013. *Material Characterization of Guadua Bamboo and the Environmental Feasibility of Structural Bamboo Products,* Massachusetts: Massachusetts Institute of Technology.

Anokye, R., Bakar, E. S., Ratnansingam, J. & Awang, K. B., 2016. Bamboo Properties and Suitability as a Replacement for Wood. *Pertanika Journal of Scholarly Research Reviews (PJSRR),* 2(1), pp. 64-80.

Anon., n.d. *Cracks in bamboo culms [Image].* [Online] Available at: https://www.permies.com/t/40855/a/24379/thumb-file1381285511484.jpg [Accessed 6 April 2017].

Arce-Villalobos, O. A., 1993. *Fundamentals of the Design of Bamboo Structures,* Eindhoven: Eindhoven University of Technology.

Awalluddin, D. et al., 2017. Mechanical properties of different bamboo species. *MATEC Web of Conferences,* Volume 138.

Baba, M. & Bhalla, S., 2009. *Shed Type Structures - Steel vs Bamboo,* Delhi: Indian Institute of Technology.

Bamboo Village Hawaii, 2004. *Bamboo Applications: Bamboo report - bamboo as a construction material.* [Online] Available at: http://www.bamboovillagehawaii.org/report.htm [Accessed 06 06 2011].

Brink, M., 2008. *Dendrocalamus asper,* Wageningen: PROTA (Plant Resources of Tropical Africa).

Bureau of Indian Standards, 2015. *National Building Code of India,* s.l.: Bureau of Indian Standards.

Burger, M. et al., 2017. *Strategies to standardise bamboo for manufacturing process chains.* Stellenbosch, South Africa, Procedia Manufacturing, pp. 330-337.

Chaowana, P., 2013. Bamboo: An Alternative Raw Material for Wood and Wood-based Composites. *Journal of Materials Science Research,* 15 March, 2(2), pp. 90-102.

City University London, 2005. [Online] Available at: http://www.staff.city.ac.uk/earthquakes/Bamboo/Bamboo.htm [Accessed 03 06 2011].

DeBoer, D. & Bareis, K., 2000. *Bamboo Building and Culture,* Bogota: s.n.

Dixon, P. et al., 2015. Comparison of the structure and flexural properties of Moso, Guadua and Tre Gai bamboo. *Construction and Building Materials,* 15 August, Volume 90, pp. 11-17.

Gnanaharan, R., 1994. *Bending Strength of Guadua Bamboo,* New Delhi: INBAR.

Gottron, J., Harries, K. & Xu, Q., 2014. Creep behaviour of bamboo. *Construction and Building Materials,* 11 June, Volume 66, pp. 79-88.

Gottron, J. M., 2014. *Structural Performance of Bamboo: Capacity under sustained loads and monotonic bending,* Pittsburgh: University of Pittsburgh.

Huang, P. et al., 2014. Non-homogeneous Thermal Properties of Bamboo. *Materials and Joints in Timber Structures.*

Huy, B. & Long, T., 2019. *A Manual for Bamboo Forest Biomass and Carbon Assessment,* Beijing: INBAR.

IBC, 2015. *International Building Code.* USA: International Code Council.

ICC Evaluation Service, 2017. *ESR-1636: Structural Bamboo Poles,* s.l.: ICC-ES.

INBAR, 2000. *Technical report 17: Bambusa balcooa.* [Online] Available at: http://www.inbar.int/publication/txt/tr17/bambusa/balcooa.htm [Accessed 27 02 2012].

ISO 22156, 2004. *Bamboo - Structural Design.* Geneva, Switzerland: ISO.

ISO 22157-1, 2004. *Bamboo - Determination of Physical and Mechanical Properties - Part 1: Requirements.* Geneva, Switzerland: ISO.

Jansen, P. C. M. & Duriyaprapan, S., 1995. Bambusa polymorpha. In: S. Dransfield & E. A. Widjaja, eds. *Plant Resources of South-East Asia No. 7: Bamboos.* Leiden: Backhuys Publisher, pp. 67-69.

Jansen, P. C. M. & Duriyaprapan, S., 1995. Dendrocalamus strictus. In: S. Dransfield & E. A. Widjaja, eds. *Plant Resources of South-East Asia No. 7: Bamboos.* Leiden: Backhuys Publisher.

Janssen, J., 1981. *Bamboo in Building Structures,* Eindhoven: Eindhoven University of Technology.

Janssen, J., 2000. *Designing and building with bamboo,* Beijing: INBAR.

Jiang, Z. et al., 2012. Sensitivity of several selected mechanical properties of Moso Bamboo to moisture content change under the Fibre Saturation Point. *BioResources,* 7(4), pp. 5048-5058.

Kaminski, S., Lawrence, A. & Trujillo, D., 2016. *Design Guide for Engineered Bahareque Housing,* Beijing: INBAR.

Kaur, P. J., Satya, S. S., Pant, K. K. & Naik, S. N., 2016. Eco-Friendly Preservation of Bamboo Species: Traditional to Modern Techniques. *BioResources,* 11(4).

Kibert, C., 2016. *Sustainable Construction - Green Building Design and Delivery.* 4th ed. Hoboken, New Jersey: John Wiley & Sons, Inc.

Kuehl, Y. & Yiping, L., 2012. *Carbon Off-setting with Bamboo,* Beijing: INBAR.

Liang, C. B. & Robinson, P., 2009. *Bamboo as a Permanent Structural Component,* London: Imperial College London.

Liese, W., 1998. *The Anatomy of Bamboo culms,* Beijing: INBAR.

Lorenzo, R., Mimendi, L. & Li, H., 2019. Digital analysis of the geometric variability of bamboo poles in bending. *MATEC Web of Conferences,* 275(01007).

Manandhar, R., Kim, J. & Kim, J., 2019. Environmental, social and economic sustainability of bamboo and bamboo-based construction materials in buildings. *Journal of Asian Architecture and Building Engineering,* 18(2), pp. 49-59.

Mbuge, D. O., 2000. *Mechanical Properties of Bamboo (Bambusa vulgaris) grown in Muguga, Kenya,* Nairobi: University of Nairobi.

Mena, J., Vera, S., Correal, J. F. & Lopez, M., 2012. Assessment of fire reaction and fire resistance of Guadua angustifolia kunth bamboo. *Construction and Building Materials,* Issue 27, pp. 60-65.

Milani, B., 2005. *Building Materials in a Green Economy: Community-based Strategies for Dematerialization,* s.l.: University of Toronto.

Murphy, R. J., Trujillo, D. & Londono, X., 2004. *Life Cycle Assessment (LCA) of a Guadua House.* Pereira, Colombia, s.n., pp. 235-244.

Naik, N. K., 2005. *Mechanical and Physico-Chemical Properties of Bamboo,* Mumbai: Indian Institute of Technology.

Nath, A., Lal, R. & Das, A., 2015. Managing woody bamboos for carbon farming and carbon trading. *Global Ecology and Conservation,* January, Volume 3, pp. 654-663.

Nugroho, N., Bahtiar, E. & Nurmadina, 2018. Grading development of Indonesian Bamboo culm: Case study on Tali Bamboo (Gigantochloa apus). *2018 World Conference on Timber Engineering.*

PlantUse English, 2016. *Gigantochloa apus,* s.l.: PlantUse English.

Quintans, K. N., 1998. *Ancient Grass, Future Natural Resource*, Beijing: INBAR.

Richard, M. J., 2013. *Assessing the Performance of Bamboo Structural Components*, Pittsburgh: University of Pittsburgh.

SANS 10160-2, 2011. *Basis of structural design and actions for buildings and industrial structures - Part 2: Self-weight and imposed loads.* Pretoria: SABS.

SANS 10163-1, 2003. *The structural use of timber Part 1: Limit-states design*, Pretoria: SABS.

SANS 10400-T, 2011. *The application of the National Building Regulations Part T: Fire protection.* 3 ed. Pretoria: SABS.

Schröder, S., 2012. *Leaching bamboo [Image].* [Online] Available at: https://static1.squarespace.com/static/53407b1de4b05fdb12d8f4a7/t/54c272cde4b0f7430cb0189e/142 2029522433/bamboo-preservation.jpg [Accessed 4 May 2017].

Shah, D., Bock, M., Mulligan, H. & Ramage, M., 2016. Thermal conductivity of engineered bamboo composites. *Journal of Material Science,* Volume 51, pp. 2991-3002.

Sharma, B., 2010. *Seismic Performance of Bamboo Structures*, Pittsburgh: University of Pittsburgh.

Solomon-Ayeh, K. A., 2010. *Use of bamboo for buildings - a sustainable, strong, versatile and economic option for the preservation of timber in Ghana*, Kumasi, Ghana: Building and Road Research Institute (BRRI).

Stamatis, D. & Gangyi, T., 2019. Deciding on Bamboo or Steel as a Building Material in Rural China: The Area X Pro0ject. *ARENA Journal of Architectural Research,* 4(1).

Tönges, C., 2002. *Shear failure [image].* [Online] Available at: http://www.conbam.info/pagesEN/properties.html [Accessed 3 June 2017].

Trujillo, D. & Lopez, L., 2016. Bamboo material characterisation. In: K. Harries & B. Sharma, eds. *Nonconventional and Vernacular Construction Materials: Characterisation, Properties and Applications.* 2nd ed. Duxford: Woodhead Publishing, pp. 491-520.

United Nations Industrial Development Organization (UNIDO), 2008. *Cottage Industry Manuals: Raw Materials and Tools for Bamboo Applications, Eastern Africa Bamboo Project,* Amsterdam: INBAR.

van der Lugt, P., van den Dobbelsteen, A. & Abrahams, R., 2003. Bamboo as a building material alternative for Western Europe? A study of the environmental performance, costs and bottlenecks of the use of bamboo (products) in Western Europe. *Journal of Bamboo and Rattan,* 2(3), pp. 205-223.

van der Lugt, P., van den Dobbelsteen, A. & Janssen, J., 2005. *An environmental, economic and practical assessment of bamboo as a building material for supporting structures.* [Online] Available at: https://www.sciencedirect.com/science/article/abs/pii/S0950061805001157 [Accessed 03 June 2011].

van der Lugt, P. & Vogtländer, J., 2015. *The Environmental Impact of Industrial Bamboo Products: Life-cycle Assessment and Carbon Sequestration*, Beijing: INBAR.

van der Lugt, P., Vogtländer, J. & Brezet, H., 2009. *Bamboo, a Sustainable Solution for Western Europe: Design Cases, LCAs and Land-use*, Beijing: INBAR.

WCGTPW, 2018. *Beaufort West Hillside Clinic completed.* [Online] Available at: https://www.westerncape.gov.za/news/beaufort-west-hillside-clinic-completed [Accessed 30 July 2020].

Widjaja, E. & Risyad, Z., 1980. Anatomical properties of some bambos utilized in Indonesia. *Bamboo Research in Asia,* May. pp. 244-249.

Xiao, Y., Zhou, Q. & Shan, B., 2010. Design and Construction of Modern Bamboo Bridges. *Journal of Bridge Engineering,* 15(5), pp. 533-541.

Yu, X., 2007. *Bamboo: Structure and Culture*, s.l.: University of Duisburg-Essen.

Chapter 5

5 AN APPROACH TO ENGINEERING MODELLING, DESIGN AND CALCULATIONS FOR BAMBOO CONSTRUCTION

5.1 Introduction

There exist a small number of engineered bamboo structures using bamboo in its natural form as whole culms. This type of construction is still considered a niche market and it is primarily dominated by architects such as Simon Velez and Jörg Stamm, both of whom operate mainly in Colombia (refer Figure 5.1 and Figure 5.2).

Figure 5.1: ZERI Pavilion (Germany) by Simon Velez (Sharma, 2010)

Figure 5.2: Pedestrian bridge (Colombia) by Jörg Stamm (Sharma, 2010)

Simon Velez worked mainly in the rural areas of Colombia. Although there were Colombian design codes in existence at the time of his work there, due to the apparent lack of enforcement of such codes in such areas, he was able to circumvent the services of a structural engineer (DeBoer & Bareis, 2000). In addition, he collaborates with his own team of workers, and mainly one species of bamboo, *Guadua angustifolia*, thereby minimising variances in the properties due to differences between species.

Jörg Stamm specialises in bridges and makes use of universities such as the Rheinisch-Westfälische Technische Hochschule (RWTH) Aachen University (Germany) and Technical University of Pereira (Colombia), for the structural engineering aspects.

Although there are numerous guidelines available for the construction of bamboo buildings and structures, these focus on construction techniques and are written predominantly by architects or building contractors. Only one guideline was found which based guidelines on engineering principles, and that was written by Jules Janssen, who is both a civil engineer and an architect, in conjunction with INBAR (International Network for Bamboo and Rattan).

There are few codes for engineering design of such structures, and these codes are limited in the information that they give, often specific to the species of bamboo found in the relevant country. This presents a dilemma for the structural engineer, as in the absence of accepted design codes, design calculations cannot be standardised, but must be calculated from first principles for every structure or based on demonstrated adequate strengths of local materials. This chapter will address this issue and attempt to provide suggestions on how to overcome these code shortcomings. Further, it will briefly cover aspects to consider for seismic design, as well as the design of connections. Lastly, this chapter will cover the use of computer software for the design of bamboo structures.

5.2   Existing Standards, Design Codes and Guidelines

One definition of a building code is "a set of rules that specify the minimum standards for constructed objects such as buildings" (Anon., 2016). In comparison, the Internal Organization for Standardization (ISO) defines standards as "Technical specifications or rules, based on consensus, and approved by a recognised standardising body" (Janssen, 2000). At an international level, this would take the form of an ISO standard (International Standards Organisation), such as ISO 22156 – Bamboo structural design. At a national level in the building industry, this takes the form of building codes of practice applicable to that country, such as SANS 10100-1 – The Structural use of concrete, the concrete design code applicable to South Africa.

In the past, there were no building codes for bamboo construction, and bamboo did not feature as a material in any national building codes. As bamboo construction has gained greater exposure and thus become more in demand, numerous countries have incorporated the use of bamboo in their design codes and standards. The location of the currently available codes and standards for structural design and construction with bamboo are shown in Table 5.1 and Figure 5.3, and discussed below (Gatóo, et al., 2014).

Table 5.1: Existing structural standards and codes for bamboo (Gatóo, et al., 2014)

Country	Standards	Codes
International	ISO 22156: Bamboo – structural design ISO 22157-1 & 2: Bamboo – determination of physical and mechanical properties	--
China	JG/T 199: Testing method for physical and mechanical properties of bamboo used in building	--
Colombia	NTC 5407: Uniones de Estructuras con *Guadua angustifolia* Kunth (Structural connections with *Guadua angustifolia* Kunth) NTC 5525: Métodos de Ensayo para Determinar las Propiedades Físicas y Mecánicas de la *Guadua angustifolia* Kunth (Methods and tests to determine the physical and mechanical properties of *Guadua angustifolia* Kunth)	*Reglamento Colombiano de Construcción Sismo resistente –* Titulo G: Estructuras de madera y estructuras de Guadua (Colombian Regulation of Seismic Resistant Construction - Chapter G12 Structures of timber and Guadua bamboo)
Ecuador	INEN 42: Bambú caña Guadua (Bamboo cane Guadua)	*Norma Ecuatoriana de la Construcción* - Chapter 17 Utilización de la Guadua Angustifolia Kunth en la Construcción (Ecuadorian Construction Standard – Chapter 17 Use of *Guadua angustifolia* Kunth in construction)
India	IS 6874: Method of tests for round bamboos IS 15912: Structural design using bamboo - code of practice	*National Building Code of India*, Section 3 Timber and Bamboo: 3B
Peru	--	*Reglamento Nacional de Edificaciones*, Section III. Code E100 - Diseño y Construcción con Bambú (National Building Regulations, Section III. Code E100 - Design and Construction with Bamboo)
USA	ASTM D5456: Standard specification for evaluation of structural composite lumber products	--

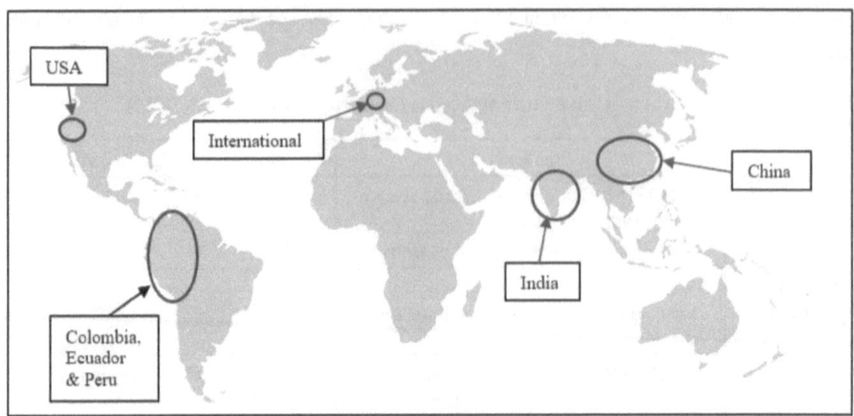

Figure 5.3: Location of Bamboo Standards and codes around the world (reproduced from Lopez, 2016)

5.2.1 Design Standards

- *International*

In 2004, the International Network for Bamboo and Rattan (INBAR) submitted building codes for bamboo to the International Standards Organisation (ISO). These were initially released as ISO 22156:2004 - Bamboo-Structural Design and ISO 22157:2004 - Determination of physical and mechanical properties of bamboo. The latest editions were reviewed and issued in 2012.

The intent of the international standard, ISO 22156, was to provide an engineering design approach while recognising traditional design and practices (Harries, et al., 2012). It follows a Limit States Design approach, with the focus on the performance of the bamboo structure. Engineering designs are to be based on calculations, with some notable exceptions. ISO permits the design to be based on "traditional" methods, if the designs are accepted within the area as general wisdom, and if the designs are implemented solely in similar situations.

A second exception is designs based on reports after natural disasters. Structures that survived are considered adequate for similar future disasters, and can thus be implemented, provided that the report is composed by acknowledged engineers, and full details are provided on the construction of those surviving structures.

Design stresses are not given in the code, but the designer is instructed to use strength and stiffness parameters derived from tests or derived from comparison with similar species. ISO further states that once the allowable stresses have been derived, a factor of safety of 2.25 should be applied in the Limit States Design.

ISO 22156 does not provide a foundation from which architects, engineers or builders can design and construct (Gatóo, et al., 2014), in that it does not provide exact structural or material properties to be used in the design process. As a result, bamboo properties, joints and connections are studied on a case-by-case basis. This approach is costly for new methods of construction, due to the effort required for each design.

The first part of the international standard, ISO 22157-1, provides the testing methodology to be used to obtain material properties for bamboo for design purposes, while the second part, ISO 22157-2, represents the laboratory manual for the test methods and data collection, described in Part 1. In the absence of properties for a specific species, this is the recommended standard to be used to derive the necessary properties.

- *China*

The Chinese standard JG/T 199 presents testing procedures to determine bamboo properties similar to those given in the ISO documents, but with different methods and parameters. It also provides a correction factor for moisture content when obtaining strength and stiffness values.

The letters "JG" indicate that this standard is a professional standard for the construction industry. The "T afterwards indicates that application of this standard is voluntary.

The main bamboo species found in China is Moso bamboo, or *Phyllostachys pubescens*, details of this species were given in Chapter 3 and 4.

- *Colombia*

The Colombian standard NTC 5407 (Norma Técnica Colombiana or Colombian Technical Standard) provides guidance on connection design when using the bamboo species *Guadua angustifolia*, while the standard NTC 5525 covers testing methods to determine the properties of this bamboo species, which is the bamboo species most commonly found in this area.

- *Ecuador*

The design standard INEN 42 describes useful aspects for bamboo construction with *Guadua angustifolia* but does not provide any design guidance. Similarly, in Colombia, this is the main bamboo species found in the area.

INEN (Instituto Ecuatoriano de Normalización) refers to the Ecuadorian Service for Standardisation, formerly known as the Ecuadorian Institute for Standardisation.

- *India*

The Indian standard provides minimum requirements for structural design and testing of bamboo (Sharma, 2010). It also covers the general principles involved in the design of structural bamboo in buildings. This standard was last revised in 2018.

- *United States of America*

The American standards provide guidance on laminated veneer bamboo only, and not construction with bamboo in its natural state or form.

5.2.2   Design codes

Janssen (2000) listed the advantages of a design code for bamboo as follows:

1) Engineering recognition - many engineers prefer designing with established materials and codes, with minimum code specifications on which to base their design decisions, as this provides a measure of reassurance in the safety of their designs.
2) Trade advantages - a recognised design code provides a means of quality control of products.
3) Increased use of bamboo - a building code leads to more social acceptance of a particular building material.

In addition, when building codes and regulations are enforced, this ensures compliance in building work. In comparison, the application of standards or norms is voluntary.

- *Colombia*

The Colombian code for seismic-resistant structures, *Reglamento Colombiano de Construcción Sismo resistente* (Colombian Regulation of Seismic Resistant Construction), includes a chapter on the use of *Guadua angustifolia* bamboo, using allowable stress design methods. It also includes element and connection design methods and guidance (Asociación Colombiana de Ingenieria Sísmica (AIS), 2010).

- *Ecuador*

The code for Ecuador, *Norma Ecuatoriana de la Construcción* (Ecuadorian Construction Standard), describes the design processes for bamboo but does not include calculations (Norma Ecuatoriana de la Construcción (NEC), 2016).

- *India*

The *National Building Code of India* addresses the structural design of bamboo in Part 3, together with timber design. It covers the use of bamboo for structures, as well as the requirements to satisfy acceptable performance. Furthermore, it provides strength limits for three classes of bamboo, representing the species recommended for construction and found in India. Examples of typical joints and connections are provided but without detailed dimensions or design capacities of the elements.

Common Indian bamboo species used for construction are *Bambusa balcooa*, *Bambusa nutans*, *Dendrocalamus giganteus*, *Dendrocalamus strictus* and *Dendrocalamus asper* (Richard, 2013). Details of these species were covered in Chapters 3 and 4.

- *Peru*

The design code for Peru, *Reglamento Nacional de Edificaciones* (National Building Regulations), covers the design and construction with bamboo for seismic-resistant structures. It also references other codes and standards, including the ISO codes.

5.2.3 Design Guidelines

INBAR has published several guidelines and material values that can be used for the preliminary sizing and design of such structures, such as the *Design Guide for Engineered Bahareque Housing* (Kaminski, et al., 2016), *Designing and Building with Bamboo* (Janssen, 2000) and *Bending Strength of Guadua Bamboo* (Gnanaharan, 1994). There are also many construction guidelines available, such as the *Manual of Bamboo Construction* (2019, in Spanish), the *Bamboo Construction Source Book* by the Hunnarshala Foundation (2013) and the *Construction Manual with Bamboo* by Stamm et al. (2014). However, care should be taken when referring to these guidelines. The details that are presented in these guidelines often apply to a specific species of bamboo as found in a particular area, and as such might not be transferable to other species from other locations. It would thus be necessary to perform design checks if such details were used in a structure, to ensure the safety and viability of the structure.

5.2.4 Future Design Codification

Little information was found regarding any future plans for the codification of bamboo design, other than the updating of the codes and standards currently in use. However, INBAR reported that the building codes developed by ISO and INBAR, namely ISO 22156 and 22157, have now been adopted as National Standards in Jamaica, the Netherlands, the Philippines and Vietnam.

5.3 Key Design Considerations for Bamboo Structures

One of the typical forms of bamboo building construction is the traditional *bahareque* construction system, also referred to as *wattle-and-daub*. This system has been amended to remove the traditional deficiencies and improve the building method with current knowledge, materials and construction techniques. *Engineered bahareque* systems typically consist of reinforced concrete foundations and upstand walls, with a timber or bamboo structural frame, and plastered walls.

The *engineered bahareque* houses are considered suitable as a low-cost construction for post-disaster contexts in developing countries (Kaminski, et al., 2016). Many of the design guidelines for these buildings can be applied to bamboo construction and design in general.

The structural characteristics of bamboo culms were discussed in Chapter 4. Based on those results, there are several key characteristics to bear in mind when designing a bamboo structure, as follows:

- Bamboo culms are weak in both compressive and tensile resistance when loaded perpendicular to the culm, i.e. shear capacity is low.

- Bamboo has greater compressive and tensile strength, and connections are simpler to construct when the loads are parallel to the fibres of the culm, i.e. along the longitudinal axis of the culm.

- It is difficult to use the full tensile capacity of the culms, as the tensile strength of the connections governs the maximum tensile capacity.

- Bamboo is more flexible and bends more than a timber member of similar outside dimensions.

- Bamboo culms are often curved, so buckling of slender sections in compression should be taken into consideration.

- If untreated and exposed to the environment, bamboo is prone to rot and insect attack (Kaminski, et al., 2016).

5.3.1 Designing for Gravity loads

Similar to conventional engineered structures, Kaminski (2016) recommends that bamboo structures be designed with a simple vertical load path to transfer the loads to the foundations. Furthermore, point loads on beams should be avoided where possible to avoid local crushing of the culms at the point of loading. He further states that for simple housing structures, the elements should all be designed as pinned elements, which simplifies the design process.

5.3.2 Designing for earthquake and wind loads

Earthquake and wind loads are similar from a design point of view in that they both exert horizontal forces on the structure which need to be transmitted to the foundations. The main differences which affect the design for these two load types are:

- Earthquake loads are proportional to the self-weight of the structure, while wind loads are not.
- Earthquake loads are cyclic, which could lead to fatigue failure at the connections.
- There is less certainty regarding the magnitude of earthquake loads, whereas wind loads are mostly well-documented and can be calculated with a reasonable level of certainty (Kaminski, et al., 2016).

Although bamboo culms perform well in seismic and wind loading conditions, due to its flexibility and lightweight nature, it also possesses several brittle failure modes which need to be considered in the design of such structures.

Kaminski (2016) recommends the following guidelines when designing for wind and seismic loading:

- For seismic loading, it is advised to keep the structure as light as possible. However, for wind loading, heavier foundations or sturdier structures would improve the stability of the structure and assist in counteracting the wind forces . In areas where both conditions prevail, the design would focus on the worst-case scenario for each element.

- Overturning of the structure should be checked. Where the overturning loads are high, all parts of the structure should function as a unit, from the roof to the foundations, as a measure to increase overturning resistance. In addition, the foundations might need to be increased in size to provide additional weight, or alternative methods of increasing gravity loads could be utilised

- A continuous lateral load path from the roof to the foundations should be provided, with the avoidance of any discontinuities.

- The structure and connections should use the inherent ductility of the bamboo culms.

- Ensure a lateral load-stability system in the structure exists through bracing or shear walls.

5.3.3   Lateral Load-stability systems

Lateral load-stability is required in structures to resist the horizontal forces from wind and seismic loading. Kaminski (2016) does not recommend the use of moment frames in bamboo design for the following reasons:

- Suitable connections with sufficient strength, stiffness and ductility are not readily available.

- Single bamboo elements might not have the necessary strength or stiffness to form portal systems. Although multiple culm bundles could be substituted, composite action between the members is difficult to achieve.

- Moment frames have little ductility and therefore, could fail without warning in a brittle manner (Kaminski, et al., 2016).

The most common lateral stability systems are braced frames or shear walls.

- *Braced frames*

In their simplest form, these triangulated frames transfer the loads to the foundations via axial loads in the elements. Kaminski (2016) recommends the use of sufficient bracing such that the tensile forces remain small (refer Figure 5.4 and Figure 5.5).

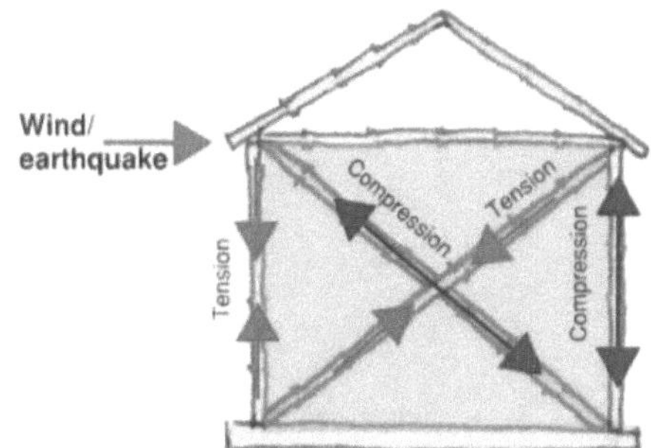

Figure 5.4: Typical bracing system for bamboo housing (Kaminski, et al., 2016)

Figure 5.5: Bracing in typical bamboo structure (Diseño en Caña de Bambú, 2016)

Similarly, the South African standard for timber frame buildings gives details and specifications for bracing, including the recommended location of the braced panels (refer Figure 5.6).

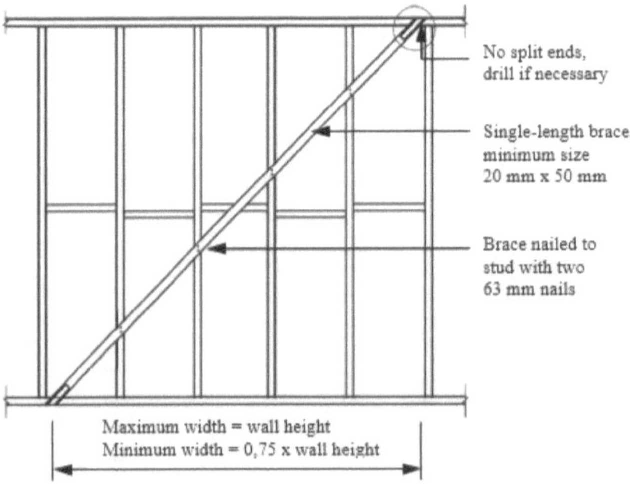

Figure 5.6: Timber bracing for timber frame buildings (SANS 10082, 2007)

- *Shear walls*

Shear walls are typically made from a continuous solid material or system which transfers the horizontal loads to the foundations by means of a shear force distributed along the length of the wall. In bamboo structures, these are often formed by the addition of plaster or cement mortar render to the wall frame, such as found in the engineered *bahareque* bamboo structures (refer Figure 5.7).

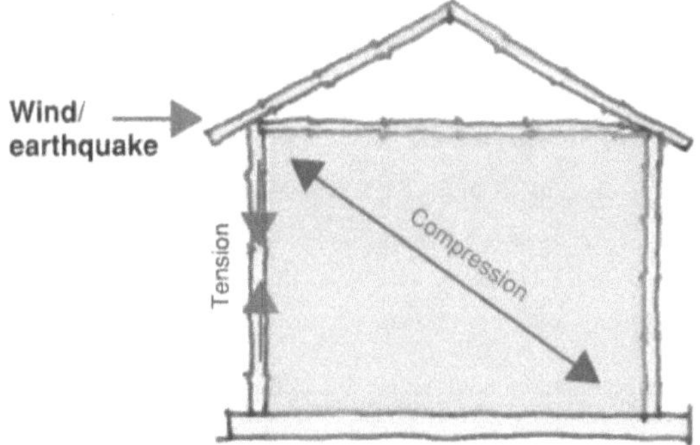

Figure 5.7: Horizontal load path for engineered *bahareque* shear walls (Kaminski, et al., 2016)

One of the advantages of engineered *bahareque* shear walls is that the structural shear wall also acts as the partition or outside wall, with no further cladding or infill material required.

5.4   Design for Various Stress Systems

5.4.1   Flexure

The National Code of India states that flexural members should be designed using the principles of beam theory. It further recommends that shear stress be checked at the smaller end of the beam (Bureau of Indian Standards, 2016).

Deflection limits are a serviceability requirement to achieve a fit-for-purpose performance of an element. This calculation check is usually done once members have been designed to meet flexural requirements, as at times, this becomes the determining criteria for a member size and not the member's flexural capacity.

Janssen (2000) recommended limitations for the allowable deflection for beams due to the flexibility of bamboo to be less than span divided by 300.

5.4.2   Shear

As mentioned in Chapter 4, bamboo fails most often in shear by splitting and cracking, due to the hollow structure and lack of cross fibres in the internodes. This type of failure occurs particularly if the loads are applied away from the culm nodes, where the bamboo culm is weakest in shear resistance.

In order to overcome this, it is recommended that bamboo members be designed so that member loads are primarily axial loads, with shear loads applied solely at the nodes. Alternatively, specialised steel connections may be designed to transfer the loads, with strengthening of the nodal points of the bamboo culm. This will be covered in more detail in Section 5.6.

Bamboo culms are at risk of longitudinal splitting, and therefore Kaminski (2017) recommends the application of a crack reduction factor of $k_{cr} = 0.5$ to the allowable shear capacity of the culm.

5.4.3   Axial Tension

Although bamboo culms have a high tensile capacity, Kaminski (2017) states that this is rarely fully utilised, as the connections tend to fail first due to shear or local crushing.

5.4.4   Axial Compression

Kaminski (2017) stated that culms in compression fail due to local crushing of the fibres in short elements, bursting of the culm wall in medium length elements and buckling in long elements. The classification of the culm lengths into short, medium or long, was based on the slenderness ratio. This approach is similar to that used for timber and steel design, except that the second moment of area is reduced by 10% in the formula for the radius of gyration, to account for the effect of taper of the bamboo, as recommended by ISO 22156 and the Indian Building Code.

Kaminski (2017) recommends the factor for the effective length of columns, as shown in Figure 5.8. The factor for the pinned condition shown on the left is the same as that for timber. However, the factor for sway mode is higher than that used in the timber code (2.1 for bamboo, 2.0 for timber). This larger factor could be to account for the greater flexibility of bamboo as compared to timber.

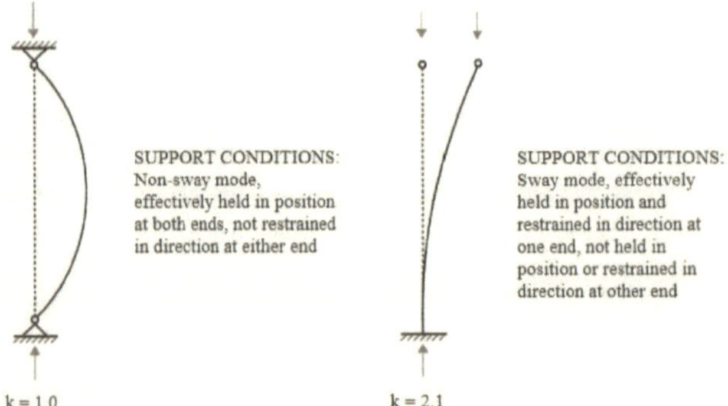

Figure 5.8: Coefficient of effective length, *k* in columns (Kaminski, et al., 2017)

Janssen (2000) recommended limitations for the allowable slenderness ratio for columns due to the flexibility of bamboo to be less than 50. These limitations are more stringent than the South African timber code, which limits the slenderness ratio for compression members to less than 52 (SANS 10163-1, 2003).

Richard (2013) mentions certain advantages of using multiple-culm columns in structures, namely: (a) simple concentric connections of framing members can be used; (b) increased axial capacities of the columns and improved resistance to lateral forces, and (c) smaller individual culm sizes can be used. He studied the ultimate axial behaviour and buckling capacities of single and multiple-culm bamboo columns, both from a theoretical and experimental perspective (refer Figure 5.9 and Figure 5.10).

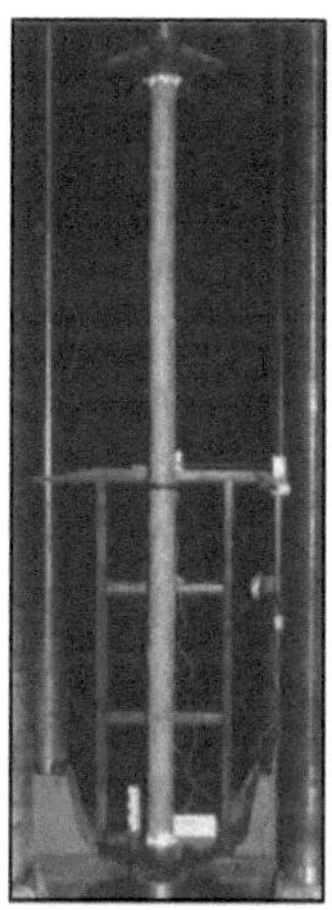

Figure 5.9: Single culm specimen prior to test (left) and after test (right) (Richard, 2013)

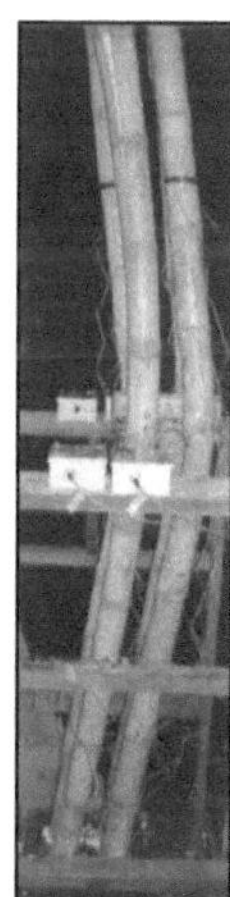

Figure 5.10: Multiple-culm specimen prior to test (left) and after test (right) (Richard, 2013)

Despite his hypothesis that connecting the culms at mid-height would result in increased section properties, his tests indicated that the multiple-culm columns distributed the load between the culms as the weaker culms failed. Furthermore, the column behaviour was a clearer representation of the sum of the section properties of the individual culms, $\sum I_{culm}$, rather than by the gross section properties, I_g, of the multiple-culm column. He concluded that multiple-culm columns would be a viable option for multi-story structures. The results of his tests are given in Table 5.2.

Table 5.2: Axial load capacities (Richard, 2013)

Specimen	Column Area $\sum A_{culm}$	Slenderness (L/r)	Critical Load P_{cr}	Critical Stress $\sigma_{cr} = P_{cr}/\sum A_{culm}$	Normalized Critical Stress $\sigma = \sigma_{cr} * (L/r)$
	(mm²)		(kN)	(MPa)	(MPa)
Single culm:					
S1	3400	97	52	15.2	1484
S2	3400	100	50	14.7	1473
S3	3800	101	96	25.0	2521
S4	3600	103	63	17.6	1809
Multiple culm:					
M1	11200	119	140	12.5	1482
M2	11700	109	159	13.6	1482
M3	13600	106	138	10.1	1071
Short culm:					
SH1	3600	71	65	18.2	1286
SH2	4600	54	131	28.6	1548
SH3	4300	27	95	22.1	607

From the results, it can be seen that a multiple-culm column has a greater axial capacity than a single-culm column, with similar overall stress.

Richard conducted similar tests on short columns to investigate the effect of slenderness ratio on the compressive strength, with the expectation that the shorter columns would indicate a greater load-carrying capacity. The results of these tests are indicated in Table 5.2.

Two of the columns tested, namely specimens SH1 and SH2, performed as expected, demonstrating a higher critical stress as compared to the single-culm and multiple-culm columns. However, the third specimen performed worse than expected, possibly due to the testing procedure.

The short column test results were inconclusive, and Richard stated that further short column tests were required.

5.5   Design of Building Systems

5.5.1   Roofs

The National Building Code of India (2015) provides some examples of roof trusses, indicating a majority of triangulated frames with pinned joints (thus no moments generated at the joints of the internal members) (refer Figure 5.11). It provides limiting dimensions, where the truss height should exceed 0.15 times the span of the truss.

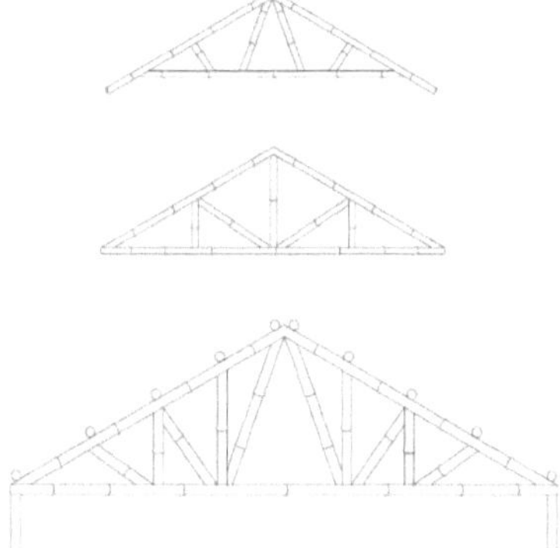

Figure 5.11: Typical roof truss configurations with pinned connections (Bureau of Indian Standards, 2016)

5.5.2   Walls

Walls protect the inhabitants from the elements and provide privacy and safety. In addition, they transfer loads from the roof to the foundations, which in turn transfer the loads to the ground below. They may also transmit horizontal seismic and wind loads to the foundations.

Walls can be constructed from whole bamboo culms and tied together with battens (refer Figure 5.12).

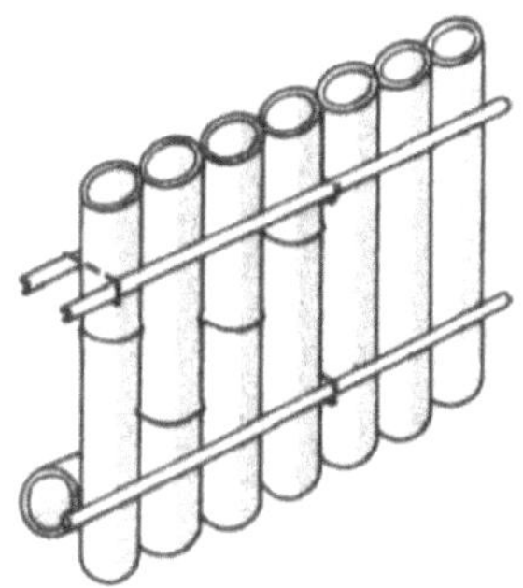

Figure 5.12: Wall from whole bamboo culms (Janssen, 1982)

Alternatively, walls in bamboo may be designed as infill structures, with the columns transmitting the forces to the foundations. These infill panels can be made from woven bamboo mats or flattened bamboo strips.

Walls can also be constructed in the *bahareque* style, with a lattice of bamboo, overlaid with mud plaster, similar to the wattle and daub technique. The wattle and daub technique uses timber poles, whereas the *bahareque* style uses bamboo culms for the main elements (Janssen, 2000), (Kaminski, et al., 2016). These walls can be considered as shear walls in the structural design, provided that the shear forces can be transmitted to the corner posts (refer Figure 5.13).

Whether a foundation wall is constructed, or whether the bamboo culms are founded in concrete bases, it is necessary to ensure a suitable connection to enable the transfer of loads (shear and axial) from the culms into the foundations. Kaminski (2016) recommends the use of steel plates bolted to the bamboo culms, or steel rods cast into the culm ends.

The use of a reinforcing bar or threaded steel rod placed inside the culm (refer Figure 5.14) provides a direct load path, without relying on the shear capacity of the culm diaphragm (Kaminski, et al., 2016). Although this detail provides some tensile capacity, it might not be suitable with large tensile loads. The tensile capacity of this connection would depend on the anchorage length of the steel rod embedded in both the culm and the foundation wall.

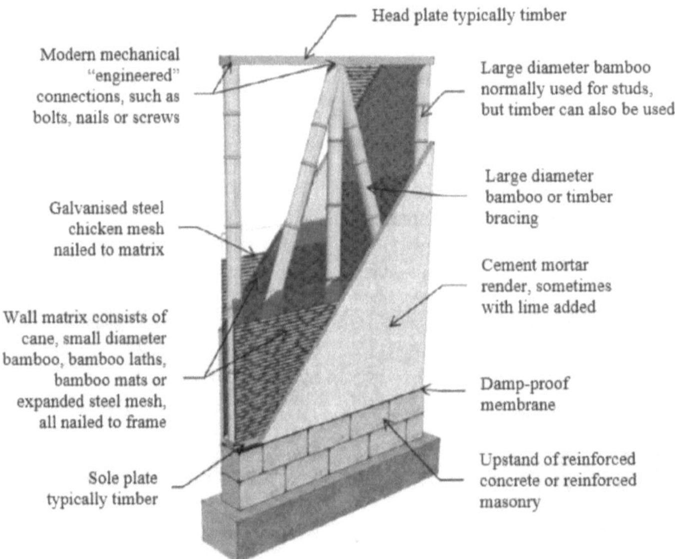

Figure 5.13: Typical bamboo lattice covered with plaster (Kaminski, et al., 2016)

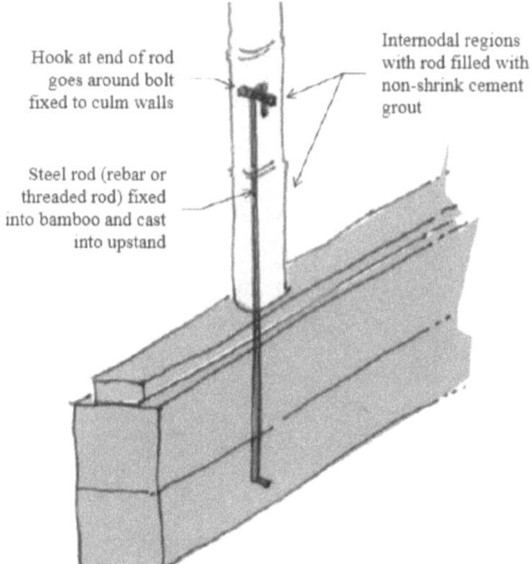

Figure 5.14: Detail of steel rod connection between bamboo culm and foundation wall (Kaminski, et al., 2016)

Where the shear forces exceed the shear capacity of the steel rod connection, a steel plate can be bolted to the culm after being cast into the foundation (refer Figure 5.15). This type of connection is also suitable where the tensile forces are high, as the steel strap provides a greater tensile resistance than the steel rod alone due to its larger area (Kaminski, et al., 2016). However, the load transfer would still be via the bearing of the fasteners on the culm wall, necessitating a bearing capacity check of the culm wall itself.

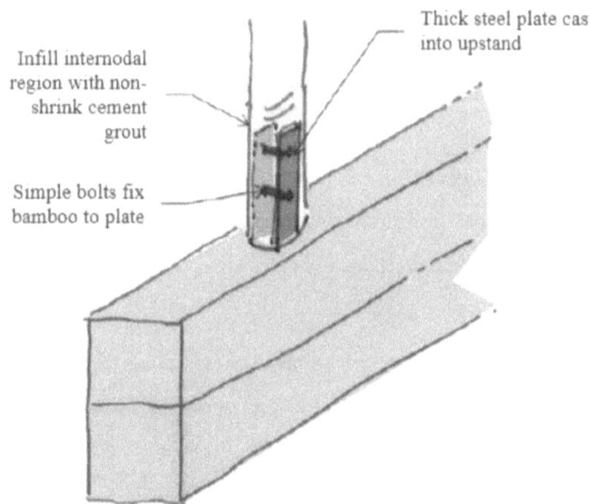

Figure 5.15: Steel plate cast into upstand and bolted to bamboo culm (Kaminski, et al., 2016)

5.5.3 Floors

Floors can be made of earth, covered with bamboo mats, or a concrete plinth or surface bed. Alternatively, the entire floor can be made from bamboo, although this would need to be elevated above the ground, in order to avoid contact with the soil (refer Figure 5.16).

Figure 5.16: Prototype of a bamboo house in Costa Rica with an elevated floor (Berg, 2016)

The South African standard for timber frame building recommends that a space be provided below the floors to allow for periodic inspection (SANS 10082, 2007). This could also be considered good practice for bamboo structures, which are vulnerable to attack from fungi and insects.

No more than two storeys are recommended by Kaminski (2016) for fire and structural reasons. However, these reasons are not explained or clarified, and so cannot be verified. This approach is similar to the South African standard for timber frame buildings, which also only provides details for structures up to two storeys in height (SANS 10082, 2007).

5.5.4 Footings

Bamboo culms should not be exposed to moisture, thus using bamboo in foundations is not advisable. Concrete foundations and wall bases are recommended for stability and to prevent the ingress of ground moisture into the culms. The National Building Code of India (2015) provides examples of three acceptable methods (refer Figure 5.17).

The method used typically in *bahareque* structures is to raise the bamboo structure above the ground using a foundation wall. The foundation wall should be able to resist the forces generated by the loading on the structure and is often constructed from masonry or reinforced concrete (refer Figure 5.18).

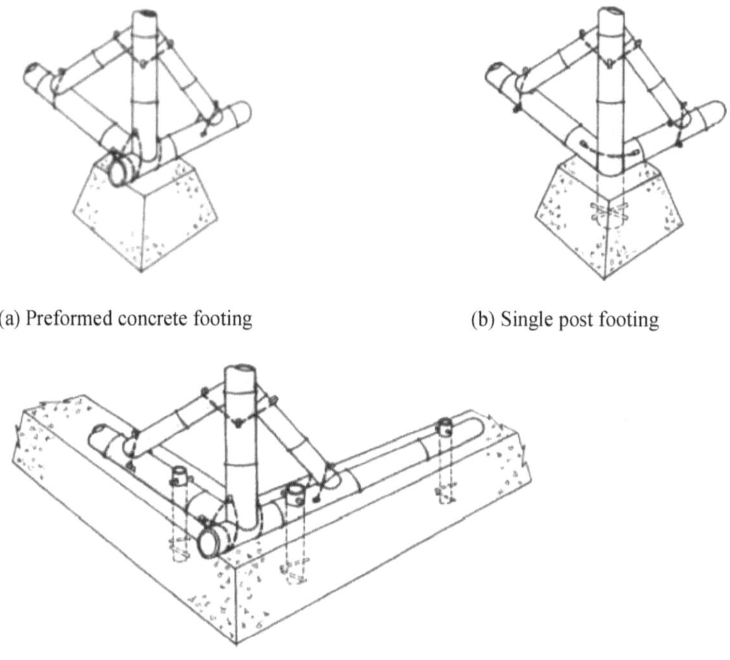

(a) Preformed concrete footing

(b) Single post footing

(c) Strip footing

Figure 5.17: Typical bamboo foundations (Bureau of Indian Standards, 2016)

It is common practice to provide a timber sole plate on the foundation wall, to provide a convenient fixing element for the bamboo structure (refer Figure 5.19). The loads from the wall elements are thus transferred via the wall plate to the foundation wall, and thus to the foundations. This practice is also recommended by the South African standard for timber frame buildings (SANS 10082, 2007).

Figure 5.18: Upstand foundation wall (Kaminski, et al., 2016)

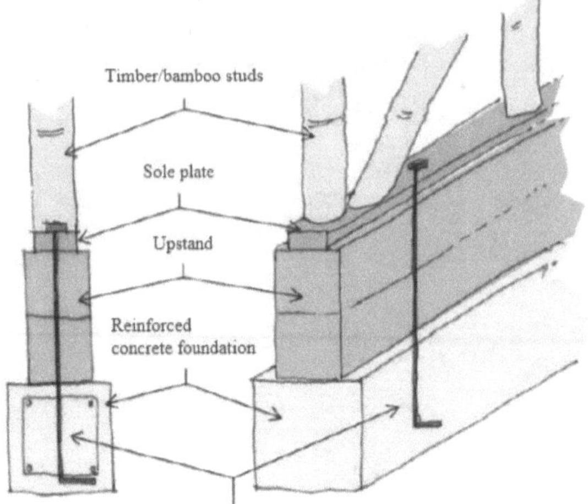

Figure 5.19: Wall plate on a foundation wall (Kaminski, et al., 2016)

5.6   Design of Connections

The design of connections and joints is a broad topic. There are many designs, techniques and details available in the literature, which will not be covered in this book. The reader is referred to the reference list for a more detailed explanation of the designs, as only a broad overview will be given here.

Despite the favourable structural and environmental properties of bamboo, one of the most significant obstacles found in bamboo construction is the connection of the members. Although timber and bamboo connections have similar issues to contend with, such as being prone to splitting, bamboo connections have issues based on their unique geometry. Bamboo culms have an elliptical, sometimes circular, cross-section, which is an inherently complex connection shape. In addition, the cross-section varies along the length, resulting in joints with connectors that are unique to each joint, or connectors that are adaptable to varying diameters and sizes, which increases costs. Also, the culms are hollow, resulting in no support in the middle of the culm to transfer the loads. Lastly, the layer of silica on the outside of the culms reduces the friction of the surface, presenting problems for any type of connection that relies wholly or partially on frictional resistance to work (Leake, et al., 2010).

Frame connections in a bamboo structure provide continuity between the culms and ensure that failure occurs in the members and not in the connections. Typical connections are lashing, pins, bolts, nails, steel wires, clamps, glued-wood fittings, and grout fills. The last two mentioned are considered to be practical for most situations ((Myers, 2013).

Traditional wood jointing techniques are generally not applicable or suitable, because of the hollow, round shape of bamboo culms. Steel joint techniques are problematic because of the large variability in the culm dimensions. Connections between members are often engineered fabricated components based on space-frame nodes (refer Figure 5.20).

Figure 5.20: Bamboo connection designed by architect Renzo Piano (Sharma, 2010)

The Colombian and Indian design codes provide some guidance on connection design. In addition, there are several empirical guidelines available in the literature. Davies (2008) suggests a few guidelines for bamboo connections, as follows:

1) Construct joint near the culm nodes, as these are strongest here due to the internal diaphragms
2) Avoid openings in the culms by making the necessary holes as small as possible. In addition, holes should be drilled and not nailed to avoid high local stress concentrations around the openings which could lead to cracking and splitting.
3) Treat the culms with preservative before construction. Shrinkage occurs during drying, so it is preferable to use the culms after the drying and shrinkage are complete.
4) Ensure that the joints fit securely together to ensure maximum transfer of loads. In practice, this often necessitates cutting the ends of the culms so that they fit flush with one another.
5) Use connections that are as durable as the structure. The structure should fail before the connections.
6) Reinforce culms under high point loads. It is preferable to avoid point load transfer perpendicularly onto the culms, as the culms have low tangential resistance. If this cannot be avoided, the area should be reinforced to avoid crushing of the culm (Davies, 2008).

After typhoons, hurricanes and earthquakes, when bamboo structures have been examined for damage, the majority of the damage has been caused by structural failure of the joints (Janssen, 2000). Reports on these disasters are useful, as they often contain details of structures that survived intact. Similar structures can then be built to withstand similar disasters in future., with reasonable assurance of their success.

5.6.1   Lashed connections

Traditional connections are simple and cheap, using natural materials that are available locally. Most common is lashing with rope, which is fast and cheap. The rope can be made from bamboo strips, rattan or other organic fibres, such as sisal or vines. It is often soaked before assembly so that the shrinkage that occurs during the drying process will tighten the lashings, and thus the connection. In some instances, the organic lashings have been replaced with plastic cords. These are solid and more reliable than natural materials but omit the shrinkage effect (Laroque, 2007). Davies (2008) illustrates two of the traditional lashing techniques (see Figure 5.21).

The disadvantage of lashed joints is that the quality in terms of stiffness and the rigidity of the connection is dependent on the skill of the worker. In addition, lashing does not use the full strength of bamboo culms and instead relies on friction for its strength. One of the dangers of using lashing as a connection method is the risk of fire. In such a situation, the lashing would fail before the culms, causing a sudden collapse of the affected structure. Lastly, complex geometries where many members are connecting at one node are challenging to construct with any stability.

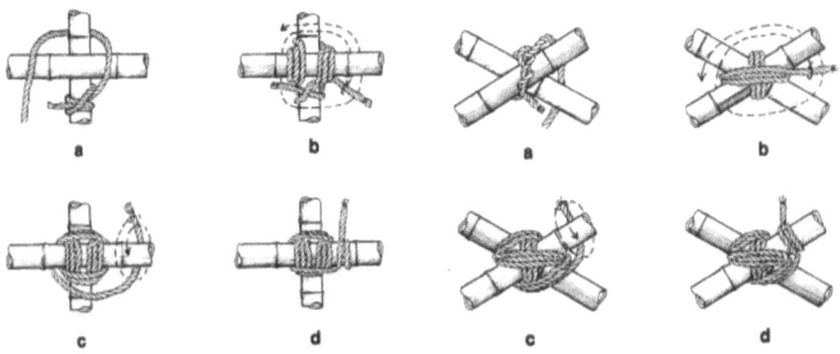

(a) Square lashing technique   (b) Diagonal lashing technique

Figure 5.21: Traditional lashing techniques (Davies, 2008)

5.6.2   Nail, bolt and pin connections

Although these connections are simple to construct, they can induce splitting in the culms, particularly in the longitudinal direction. Drilling for the required holes can crack and damage the culms, as well as reduce the culm strength at the holes ((Myers, 2013). Thus, this type of connection is not recommended.

In the situation where a longer culm is required than is available, or when a stronger beam or column is required, this can be handled by tying or bolting two culms together, as indicated in Figure 5.22.

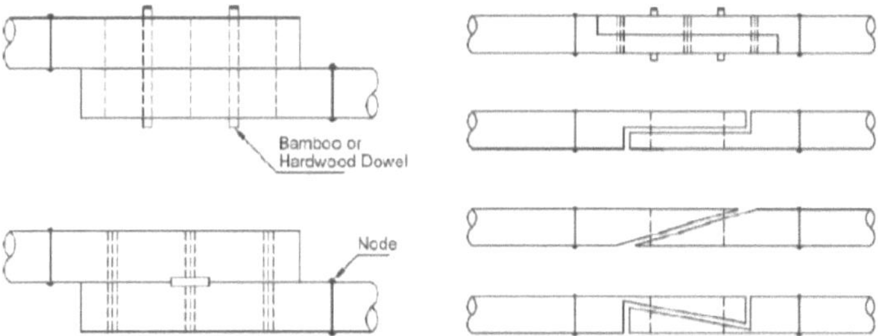

Figure 5.22: Full lapped (left) and half-lapped joints (right) (Bureau of Indian Standards, 2016)

5.6.3 Glued-wood fitting

This type of connection has a low cost, uses easily accessible materials and tools, is versatile and simple, and increases the area of the culm at the connection (refer Figure 5.23).

The wood insert is fitted inside the culm and glued in place. The forces are transferred from the culm through the wood fitting into the next member or element. The shear stresses are redistributed, and the bending stresses are reduced in the connection area. In addition, the culms are protected by the closure of the end, preventing insects from entering (Myers, 2013).

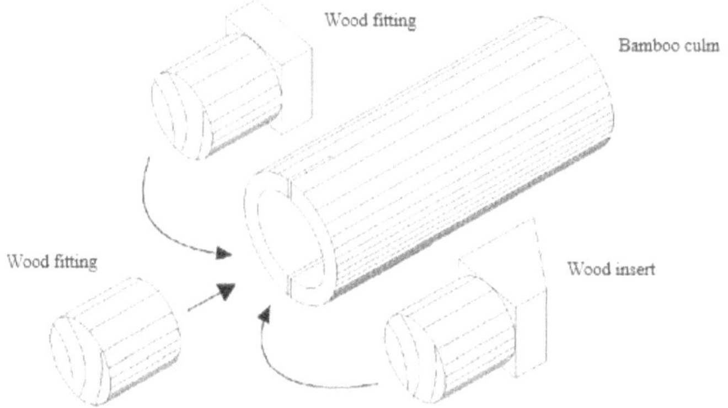

Figure 5.23: Glued-wood fitting Connection (Myers, 2013)

These are also known as Plug-in connections and are often used for larger diameter culms, using bolts. The drilling of the bolt holes can lead to end shear failure of the culms or longitudinal cracking. In addition, if the culm shrinks after construction, the connections will lose stability. For this reason, these connections are often strengthened with additional lashing or bolts (see Figure 5.24).

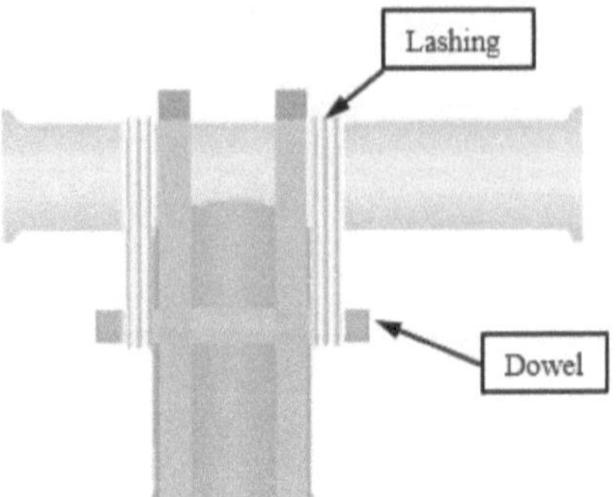

Figure 5.24: Typical dowel connection with additional lashing (Janssen, 2000)

5.6.4 Grout-filled connection

The cement grout in the connection transfers forces at the culm end similar to that of the wood-filled connection (refer Figure 5.25).

This type of connection is suitable for trusses and foundation connections, although it might not be practical elsewhere in a structure (Myers, 2013). One disadvantage of this connection type is the moisture in the grout, which could cause the culm to swell at placement. When the grout and the culm dry, this could lead to shrinking, with resultant cracking or splitting of the culm.

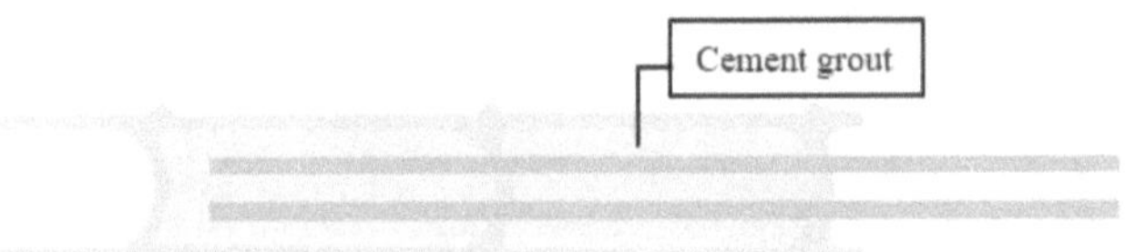

Figure 5.25: Grout-filled connection with steel rebar (Janssen, 2000)

5.6.5 Steel plate and bolts

As a result of past problems, innovative connections have been developed, using steel elements to transfer the stresses (refer Figure 5.26).

The fact that the bamboo culm is irregular in diameter creates problems regarding standard end connections. This dilemma has been resolved by several engineers and architects, one of which is shown in Figure 5.27.

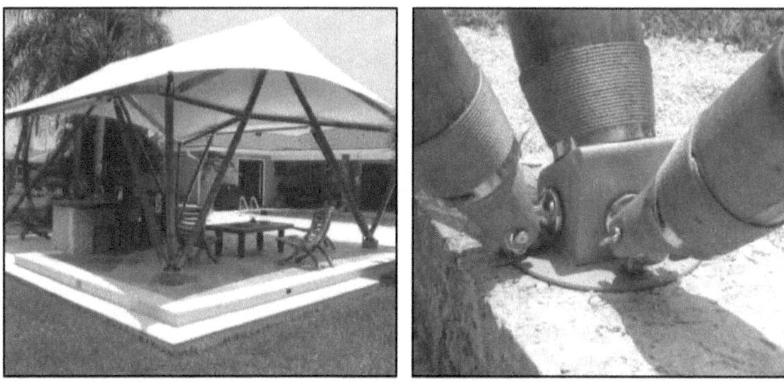

Figure 5.26: Bamboo pavilion and steel connections by Koolbamboo (Yu, 2007)

Figure 5.27: CONBAM building system developed by Christoph Tönges (Vogtländer, et al., 2010)

An alternative is to construct gusset plates from timber or plywood and fix the plates to the bamboo culms with bolts or dowels (refer Figure 5.28). Although drilling into the culms can reduce their shear capacity, the use of the gusset plates provides greater flexibility for the connections, to the extent that a mass-produced gusset plate can be used for bamboo culms of varying diameter.

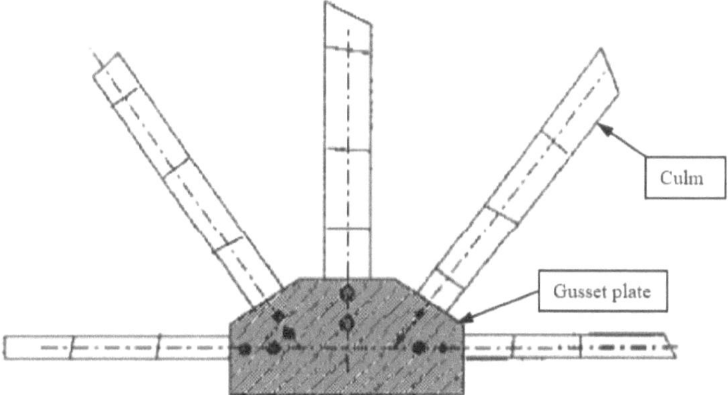

Figure 5.28: Gusset Plate Design (Davies, 2008)

Hogan (2009) proposed a reinforcing rod embedded in mortar in the end of the culm, which would then be welded to a steel gusset plate (refer Figure 5.29). Preliminary testing of this connection gave a pull-out strength of 20 kN

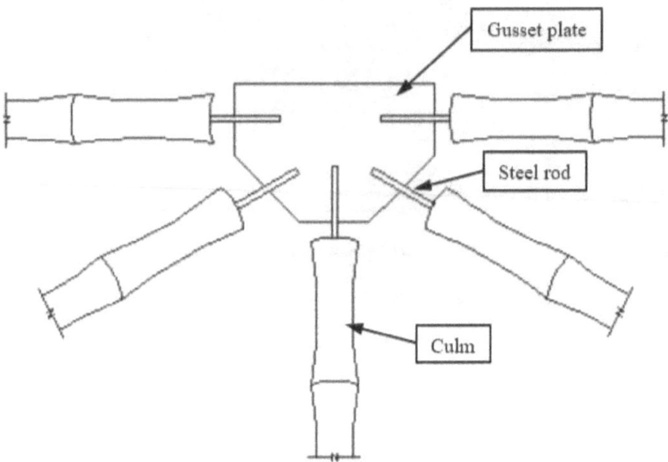

Figure 5.29: Steel gusset plate connection (Hogan & Archer, 2009)

Liang (2009) provided a useful summary of the advantages and disadvantages of several joinery methods (refer Table 5.3).

Table 5.3: Advantages and disadvantages of various joinery methods (Liang & Robinson, 2009)

Method		Advantages	Disadvantages
Mortice and Tenon Joinery		- Simple: insertion of smaller bamboo (tenon) into a hole in a larger bamboo (mortice) - Cheap and light: no other material required	- Diameter of the mortice bamboo must be large enough - Does not constrain a tensile force along the x-direction of smaller bamboo (tenon) - Involves drilling a large hole and risks damage to the structural integrity of the bamboo culm
Lashing		- Does not involve drilling and does not affect the structural integrity of the bamboo - Able to join bamboo in any direction	- Strength of the joint is dependent on the strength of the lashing material and how tightly it is lashed around bamboo - Will fail if lashing disintegrates or loosens over time
Steel Bolt and Nut Joinery		- Simple and strong - Able to join many bamboo culms together in any direction - Circular bolt hole enables stresses to be evenly distributed around the bolt hole	- Involves drilling a large hole and affects the structural integrity of the bamboo culm to a certain extent - Requires at least two bolts in a bamboo culm to constrain it

Table 5.3 (continued):

Steel Plate Inserts and Bolts		- Usage of steel inserts and bolts result in a secure and robust joint that is constrained in every direction	- Expensive and heavy due to the usage of steel - Involves drilling of a circular bolt hole and a large rectangular hole for the plate, which damages the structural integrity of the bamboo culm to a considerable extent - Diameter of bamboo culm must be large enough, and the walls must be thick enough to hold the steel plate in place
Internal Steel Plate and Bolts (triangulated)		- Strong, stiff and secure - Able to join many bamboo culms together in every direction - Triangulated metal plates help to relieve stress around joint	- Heavy - Involves drilling of bolt holes and the cutting of a slot for the steel plate, damages the structural integrity of the bamboo culm - Presence of steel plate internally increases the risk of failure by rupture - Requires bamboo culm to be thick to hold steel plate in place
External Steel Plate and Bolts (triangulated)		- Strong, stiff and secure - Able to join many bamboo culms together in every direction - Triangulated metal plates help to relieve stress around joint	- Heavy - Involves drilling of bolt holes and damages the structural integrity of the bamboo culm

5.7   Seismic Design and Construction Guidelines

The seismic force generated by an earthquake is a product of the mass of the structure, and the ground acceleration, which is a property of the earthquake itself. The objective of the seismic design of buildings is to prevent the collapse of the building, not to prevent the movement due to the seismic force.

As bamboo is lightweight, resulting in a low building weight, the seismic forces generated from certain ground accelerations would be less than those for a conventional brick and mortar house. In addition, bamboo has high elasticity, and is thus more ductile, and thus can resist the seismic forces better than conventional buildings.

In April 1991, 20 bamboo houses in Costa Rica survived a 7.5 Richter earthquake near the epicentre, while the surrounding masonry structures sustained considerable damage (DeBoer & Bareis, 2000). The first explanation is the light weight of the bamboo structures, which attract smaller lateral seismic forces. Secondly, bamboo is more flexible, with both the culms as well as the typical lashed connections providing some "give" at the connections, allowing greater flexibility.

In February 2006, in the northeast Indian hill region of Sikkim, newly constructed masonry and concrete structures were damaged in an earthquake measuring 5.7 on the Richter scale, while the traditional bamboo frame structures, built with little or no engineering or structural design, experienced little or no damage (refer Figure 5.30, Figure 5.31 and Figure 5.32:) (Sharma, 2010).

Figure 5.30: Damaged masonry building following the Sikkim earthquake, India (Sharma, 2010)

In rural areas of Sikkim, many buildings are constructed of stone masonry with a mud mortar overlay. These typically do not incorporate any seismic resistant features. As a result, these buildings experienced cracks near the corners and at openings during the earthquake, as shown in Figure 5.30.

Figure 5.31: Damage to a reinforced concrete building following the Sikkim earthquake (Sharma, 2010)

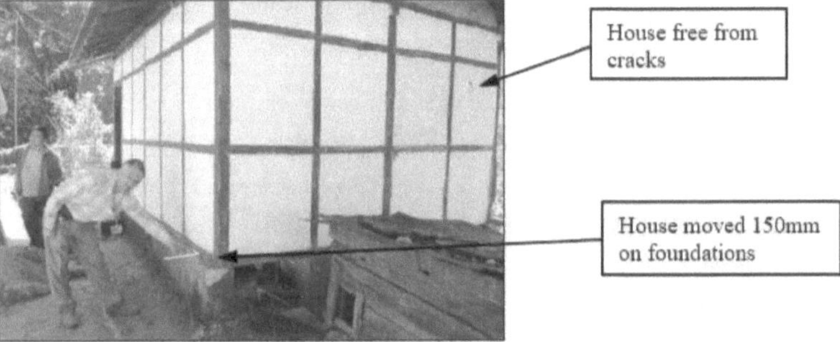

Figure 5.32: Undamaged bamboo frame construction building following the Sikkim earthquake (Sharma, 2010)

The Indian Plywood Industries Research and Training Institute (IPIRTI) developed and constructed earthquake-resistant prototype bamboo houses with the following main features (City University London, 2005):

1) Whole bamboo culms as columns and trussed rafters every 1.2m as the main load-bearing elements.

2) Infill wall panels formed from split bamboo grids, chicken-wire mesh and cement mortar plaster, where the walls are designed to provide lateral stability.

3) Application of preservative – Boron to the grids and trusses, and Creosote oil to the columns.

4) Bamboo mat board gussets and mild steel bolts for load-bearing joints in the roof structure.

5) Bamboo mat corrugated sheets as roof cladding, and bamboo mat boards for the walls, doors and window shutters (City University London, 2005).

A typical house built according to this system is shown in Figure 5.33.

Figure 5.33: Prototype of a Bamboo house by IPIRTI (City University London, 2005)

Details of the construction method can be found in the referenced articles (City University London, 2005).

In 2004, the Indian Plywood Industries Research and Training Institute (IPIRTI) in Bangalore, together with TRADA (UK) tested a full-scale prefabricated bamboo house in a series of shake table tests (refer Figure 5.34). The building was 2.7m square in plan and weighed 2636 kg. It was subjected to ground motions based on the 1995 Kobe earthquake, which had a magnitude of 7.2 on the Richter scale. Although detailed results were unavailable at the time of writing, the house reportedly showed not cracks or damage after the testing (Sharma, 2010).

Figure 5.34: IPIRTI (Bangalore) and TRADA (UK) full-scale bamboo house (Sharma, 2010)

In 2010, Sharma created a theoretical model, based on St. Joseph's School in Mungpoo, India (refer Figure 5.35). This school was a single storey building with four classrooms. It was constructed with reinforced concrete foundations (Figure 5.35d), with rubble infill below the floor (Figure 5.35c). The columns were formed from multiple culms and bolted together to form the primary frame (Figure 5.35e).

(a) Front elevation

(b) End elevation

(c) Reinforced concrete grade beam foundation with rubble infill

(d) Reinforced concrete column plinths with four bamboo culms forming one column

(e) Bolted connection used to connect primary framing

Figure 5.35: Details of St. Joseph's School, Mungpoo, India (Sharma, 2010)

After applying seismic forces to the model, as derived from actual reports of the seismic activity in that area, the results indicated that the model performed within the required limits for safety.

It can be concluded that, provided the layout recommendations for buildings in seismic areas are adhered to, bamboo construction is viable for structures located in seismic zones.

5.8 Modelling on Computer Software

A large portion of structural design is done on computers, using either spreadsheets or structural design software. However, these design approaches typically assume that a design element consists of a uniform material throughout, although this is not always the case.

Many materials and their associated properties can be modelled simply, with straight lines or flat plates. However, bamboo is not a uniform material. It tapers from along its length, has irregularly spaced nodes, and an elliptical cross-section. In addition, many culms also have a slight curve. This nonuniformity in shape creates a dilemma for the engineer in accurately representing the unique properties of bamboo in a computer model.

When modelling a functionally graded material such as bamboo on Finite Element Analysis (FEM), it is necessary to take the variation of material properties into account, if accuracy of the results is required. In traditional finite element formulation, the properties are assumed to be constant with each element that is modelled.

If the variation in properties is modelled using piece-wise constant elements, this generates an artificially discontinuous stress field. This is not an accurate simulation and will produce skewed results.

To combat this, Silva et al. (2008) proposed the use of graded elements incorporating actual material properties at integration points. He approximated the shape of a bamboo cell, neglecting the taper of the wall, and estimated the variation of Young's modulus through the bamboo thickness (refer Figure 5.36). The generated mesh for the FEM analysis is shown in Figure 5.37. He concluded that the extra effort required to model the graded variation is justified when designing for local stresses near supports, pin connections or holes. However, the effort involved to design to this level of detail would have to be weighed up against the usefulness of the information. In many instances, it is most likely that such fine design detail would be superfluous.

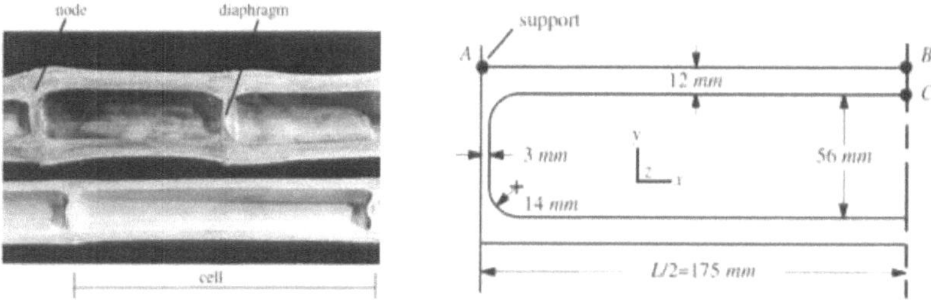

Figure 5.36: (left) Cross-section of bamboo culm; (right) section view of one-half of a cell for finite element mesh (Silva, et al., 2008)

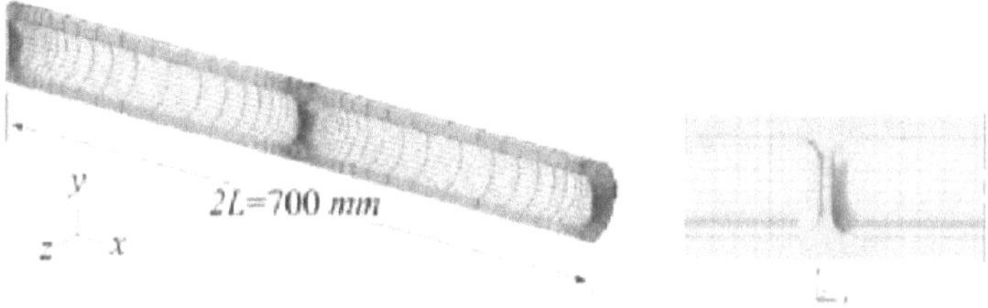

Figure 5.37: Finite element mesh of bamboo (Silva, et al., 2008)

In 2009, a project proposal was put forward by EngINdia, a partnership between the University of Cambridge, Massachusetts Institute of Technology and the Indian Institute of Technology in Bombay. The project was to design and analyse a three-dimensional model of a house frame made from bamboo, to promote appropriate and sustainable engineering solutions in developing areas. Liang (2009) lists general guidelines for how a bamboo house framework should be constructed, with emphasis on aspects like lengths of bamboo culms available.

An idealised framework was modelled in three dimensions in ProEngineer, a Finite Element Analysis software program. Each culm was modelled as a constant, uniform hollow cylinder, with material properties derived from research, and mechanical properties such as the moment of inertia calculated from average diameters. The walls were modelled as shells with a thickness of 50 mm, in order to apply wind loads over the panels (refer Figure 5.38).

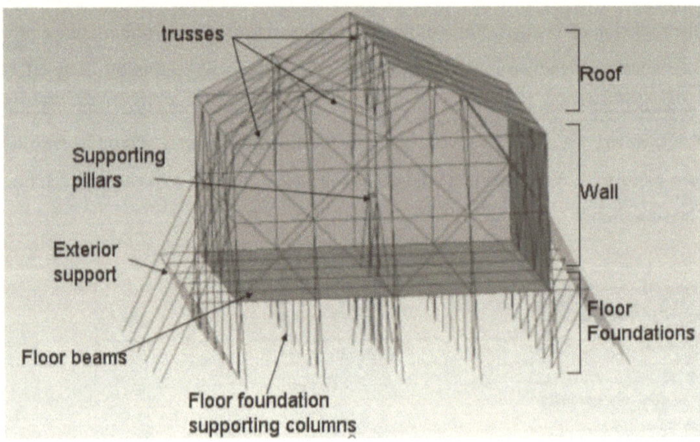

Figure 5.38: Idealised house framework (Liang & Robinson, 2009)

The expected design loads were calculated and applied, and the resulting maximum forces and stresses in each element/member noted. These results were then applied to models of the individual culms, where the nodes and bolt holes were modelled (refer Figure 5.39).

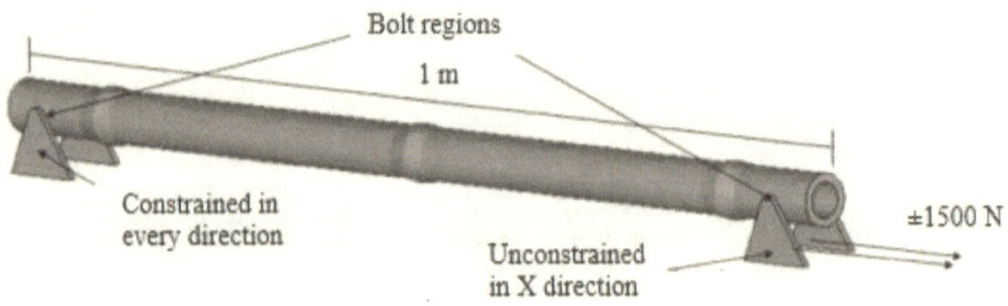

Figure 5.39: Model of culm in 3D (Liang & Robinson, 2009)

The stress distributions were noted and compared with the maximum allowable values of the culms (refer Figure 5.40). From this, it could be determined if the proposed culm had the required strength, or if additional strengthening of the members would be required.

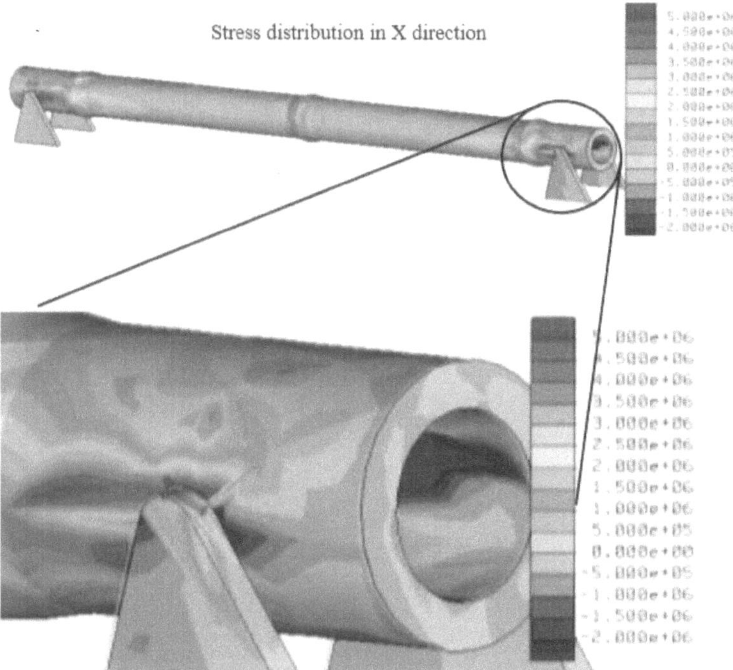

Figure 5.40: Longitudinal stress distribution on individual culm model (Liang & Robinson, 2009)

Liang (2009) listed some limitations with his approach to the computational modelling of the bamboo framework. Firstly, in the idealised model, the orthotropic properties of the culms were neglected, and only the isotropic properties were modelled. This led to an over-estimation of the deflections and strains. In addition, the idealised model assumed that one culm would be sufficient for each member. If it were found that more than one culm was necessary, the properties in the idealised model would have had to be adjusted accordingly, and the program rerun with the new values.

Secondly, the values that Liang used for input of the material and mechanical properties in his model were based on values found in the literature. Without testing of samples of the actual bamboo culms to be used in construction, the values can be either over-or under-estimated. Thus, it appears that this method is suitable for the initial design of a framework, but actual material properties would be necessary for final design and verification.

Sharma (2010) modelled a bamboo portal frame in two dimensions (refer Figure 5.41). The geometric and material properties of the individual bamboo culms were modelled using beam elements with appropriate section properties, based on the material properties of *Phyllostachys aurea*. A lateral load was applied to the top left-hand column joints (nodes 2 and 15 in Figure 5.41). The theoretical results from this indicated good agreement when compared with the actual physical experimental results (refer Figure 5.42). Thus, Sharma concluded that the theoretical model had captured the basic behaviour of the proposed portal frame.

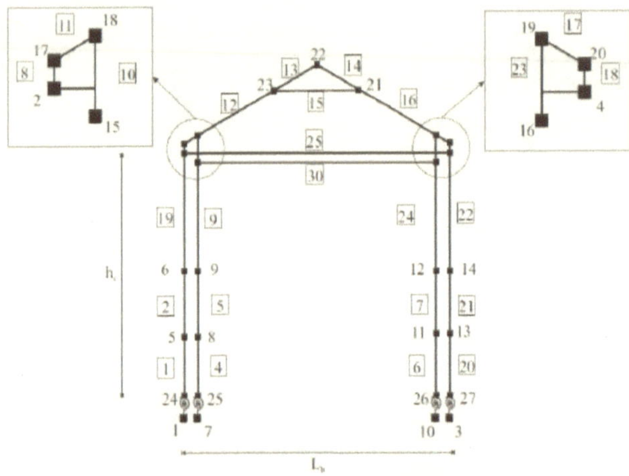

Figure 5.41: Two-dimensional model with node and member (noted in squares) numbering (Sharma, 2010)

Figure 5.42: Experimental model (Sharma, 2010)

Bhalla (2008) presented a detailed structural analysis and design of a typical industrial shed structure in accordance with the Indian standard codes of practice (refer Figure 5.43). He proposed the replacement of the conventional steel structure with a combination of bamboo and concrete, referred to as "Bambcrete" columns (refer Figure 5.44). In addition, the steel roof trusses were replaced with bamboo tied arches (refer Figure 5.45).

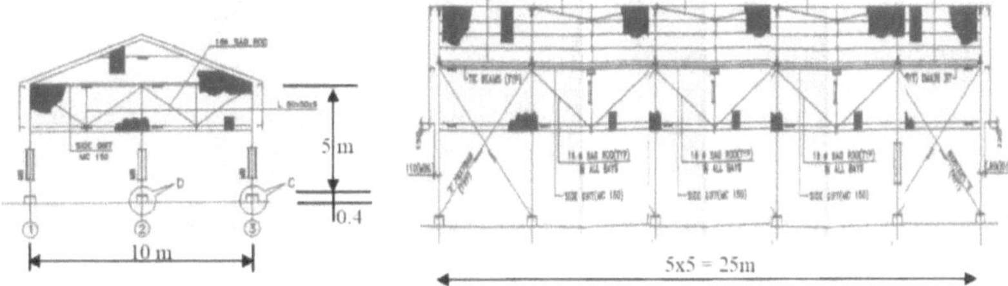

Figure 5.43: Conventional shed structure made of steel (Baba & Bhalla, 2009)

Figure 5.44: Bambcrete column (Bhalla, et al., 2008)

Figure 5.45: Bamboo bow beam for supporting roof (Bhalla, et al., 2008)

The calculations and designs were based on the use of the species *Dendrocalamus giganteus*, with a tensile strength of 121.5 MPa and compressive strength of 55.5 MPa. In addition, the culms were assumed to have an external diameter of 40 mm and a wall thickness of 10 mm. In the calculations, a factor of safety of 4 was allowed, due to the expected considerable variation in strength characteristics from the material.

Two alternative designs were presented for the connection of the bamboo columns with concrete pedestals (refer Figure 5.46 and Figure 5.47).

Bhalla concluded that the proposed structure could withstand the required loads and would cost less than the conventional structures using concrete and steel.

Myers (2013) produced calculations and designs as an example of a structurally engineered bamboo house. He used a traditional circular East African house, known as an Amhara bamboo house, with a diameter of eight meters (refer Figure 5.48).

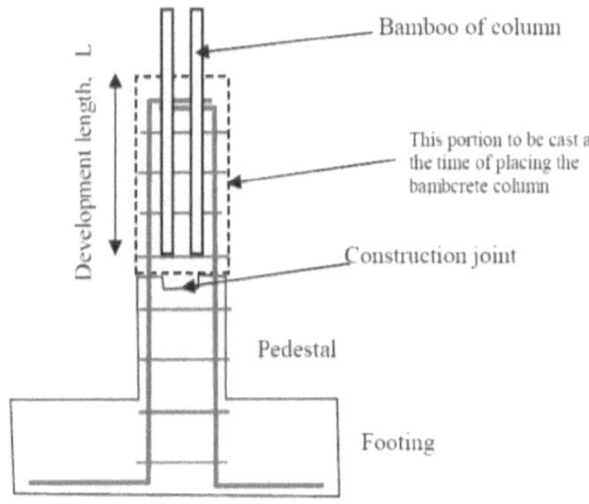

Figure 5.46: Type 1 base connection (Bhalla, et al., 2008)

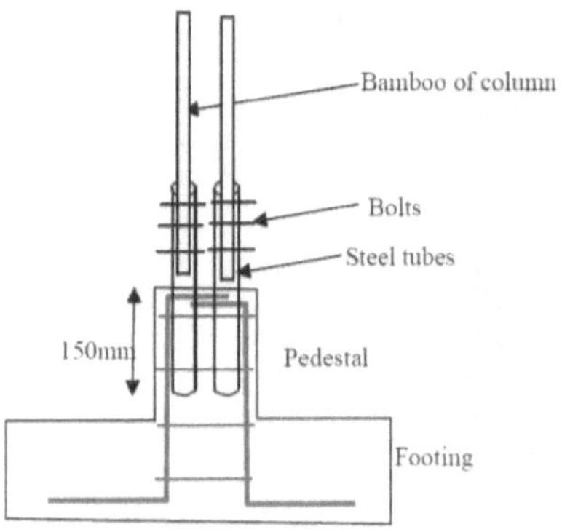

Figure 5.47: Type 2 base connection (Bhalla, et al., 2008)

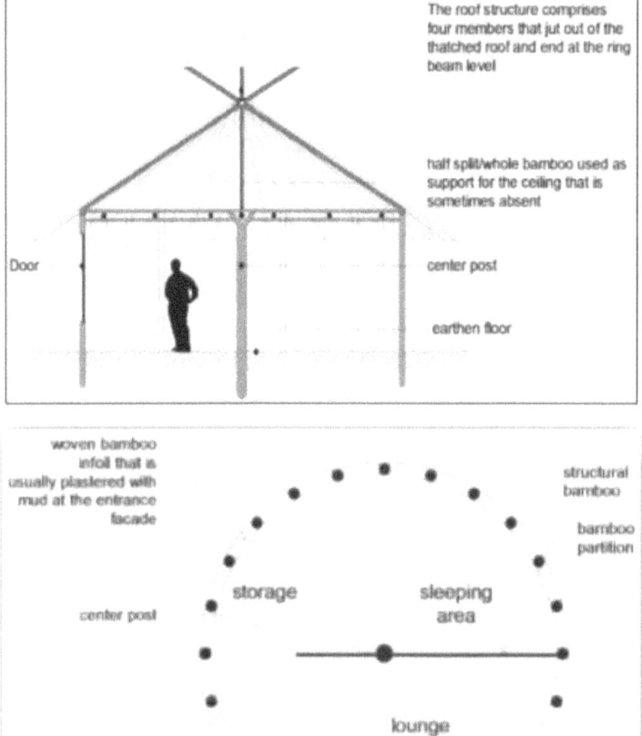

Figure 5.48: Traditional Amhara house in East Africa (Myers, 2013)

Although the central column is traditionally formed from one central timber column and the outer columns are anchored in a rammed earth foundation, for his example, Myers used an interior column formed from four bamboo culms and a concrete pedestal or foundation. His design was based on the bamboo species *Oxytenanthera abyssinica*, using two sizes, namely 60 mm or 100 mm in diameter, with wall thicknesses of 15 mm or 35 mm respectively (refer Figure 5.49, measured in centimetres).

Myers designed trusses to support the roof loads, radiating from the central column. These were designed to handle mostly tensile forces, in order to utilise bamboo's tensile strength (refer Figure 5.50 and Figure 5.51).

The results from the computer analysis indicated that the structure was safe, and Myers concluded that the computer model was an accurate representation of the actual houses built.

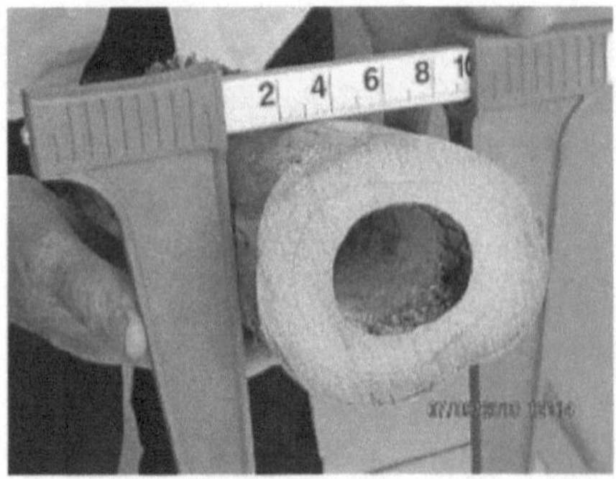

Figure 5.49: Measurement of *Oxytenanthera abyssinica* (Myers, 2013)

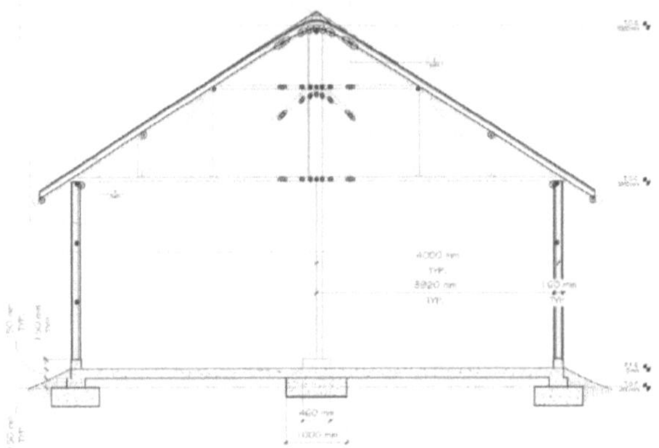

Figure 5.50: Typical cross-section of design example (Myers, 2013)

Figure 5.51: Three-dimensional model of design example (Myers, 2013)

5.9 Conclusions

There are currently no South African codes or guidelines for the design or construction of bamboo structures. Engineers are referred to either the South African timber design standard (SANS 10163:2003 Parts 1 & 2) or the gum pole design standard (SANS 754:2015 and SANS 457:2008 Parts 2 and 3). Alternatively, the international ISO design standards 22156 and 22157 could be used for design purposes, with the appropriate material properties obtained from the bamboo species considered. Although the design codes from India, Peru, Ecuador and Colombia give properties based on specific bamboo species, the calculation methods and approaches can be applied to other species.

If no reliable material properties of a specific bamboo species are available, several recommended values and approaches have been put forward, to do preliminary sizing of the bamboo members. Once testing has been done, detailed design calculations can be used to refine and optimise the design.

Due to its light weight and flexibility, it was shown that bamboo is suitable as a construction material in seismic zones.

Finally, recommendations were given for calculation of member sizes, as well as suggestions on how to model a variable material such as bamboo on a computer.

Although it has theoretically been shown that bamboo is a favourable material to use in the construction of buildings as a primary structural member, the question of how bamboo performs in actual construction remains. This will be addressed in the next chapter.

5.10 References

Anon., 2016. *Building Code*. [Online] Available at: http://en.wikipedia.org/wiki/Building_code [Accessed 20 January 2016].

Asociación Colombiana de Ingenieria Sísmica (AIS), 2010. *NSR-10: Reglamento Colombiano de construcción sismo resistente. Titulo G: Estructuras de madera y estructuras de guadua,* Bogota: AIS.

Baba, M. & Bhalla, S., 2009. *Shed Type Structures - Steel vs Bamboo,* Delhi: Indian Institute of Technology.

Berg, N., 2016. *Blooming Bamboo: A cheap prototype for disaster proof housing [image].* [Online] Available at: http://www.ecobuildingpulse.com/projects/blooming-bamboo-house-a-cheap-prototype-for-disaster-proof-housing [Accessed 15 May 2017].

Bhalla, S., Gupta, S., Sudhaker, P. & Suresh, R., 2008. Bamboo as a green alternative to concrete and steel for modern structures. *Journal of Environmental Research and Development,* 3(2).

Bureau of Indian Standards, 2016. *National Building Code of India,* s.l.: Bureau of Indian Standards.

City University London, 2005. [Online] Available at: http://www.staf.city.ac.uk/earthquakes/Bamboo/Bamboo.htm [Accessed 03 06 2011].

Davies, C., 2008. *Bamboo Connections,* Bath: University of Bath.

DeBoer, D. & Bareis, K., 2000. *Bamboo Building and Culture,* Bogota: s.n.

Diseño en Caña de Bambú, 2016. *Bamboo construction [image].* [Online] Available at: https://arteybambu.wordpress.com/sistema-constructivo/ [Accessed 15 July 2017].

Gatóo, A. et al., 2014. Sustainable structures: bamboo standards and building codes. *Engineering Sustainability,* 167(ES5), pp. 189-196.

Gnanaharan, R., 1994. *Bending Strength of Guadua Bamboo: Comparison of Different Testing Procedures,* New Delhi: KFRI & IDRC.

Harries, K. A., Sharma, B. & Richard, M., 2012. Structural Use of Full Culm Bamboo: The Path to Standardization. *International Journal of Architecture, Engineering and Construction,* June, 1(2), pp. 66-75.

Hogan, L. & Archer, G. C., 2009. *Development of Long Span Bamboo Trusses,* San Luis: California Polytechnic.

ISO 22156:2004, 2004. *Bamboo - Structural Design.* Geneva, Switzerland: ISO.

ISO 22157-1:2004, 2004. *Bamboo - Determination of Physical and Mechanical Properties - Part 1: Requirements.* Geneva, Switzerland: ISO.

Janssen, J., 1982. *A Series of Articles on the Uses of Bamboo in Building Construction,* London: Intermediate Technology Publications.

Janssen, J., 2000. *Designing and building with bamboo,* Beijing: INBAR.

Kaminski, S., Lawrence, A. & Trujillo, D., 2016. *Design Guide for Engineered Bahareque Housing,* Beijing: INBAR.

Kaminski, S. et al., 2017. Structural use of bamboo. Part 4: Element design equations. *The Structural Engineer,* 01 March, pp. 24-27.

Laroque, P., 2007. *Design of a Low-Cost Bamboo Footbridge,* Massachusetts: Massachusetts Institute of Technology.

Leake, G., Toole, K., Divis, P. & Torres-Sanchez, C., 2010. *Bamboo as a solution for low-cost housing and storage in Pabal (India).* s.l., University of Strathclyde.

Liang, C. B. & Robinson, P., 2009. *Bamboo as a Permanent Structural Component,* London: Imperial College London.

Lopez, L. F., 2016. *Bamboo Standards.* Damyang, South Korea, World Bamboo Congress.

Myers, E. T., 2013. *Structural Bamboo Design in East Africa,* Kansas: Kansas State University.

Norma Ecuatoriana de la Construcción (NEC), 2016. *Utilización de la Guadua Angustifolia Kunth en la Construcción,* s.l.: NEC.

Richard, M. J., 2013. *Assessing the Performance of Bamboo Structural Components,* Pittsburgh: University of Pittsburgh.

SANS 10082, 2007. *Timber Frame Buildings,* Pretoria: SABS.

SANS 10160-4, 2011. *Basis of structural design and actions for buildings and industrial structures - Part 4: Seismic actions and general requirements for buildings.* Pretoria: SABS.

SANS 10163-1, 2003. *The structural use of timber Part 1: Limit-states design,* Pretoria: SABS.

Sharma, B., 2010. *Seismic Performance of Bamboo Structures,* Pittsburgh: University of Pittsburgh.

Silva, E. C. N., Walters, M. C. & Paulino, G. H., 2008. *Modelling Bamboo as a Functionally Graded Material,* s.l.: American Institute of Physics.

Vogtländer, J., van der Lugt, P. & Brezet, H., 2010. The sustainability of bamboo products for local and Western European applications, LCAs and land-use. *Journal of Cleaner Production,* 3 June, Volume 18, pp. 1260-1269.

Yu, X., 2007. *Bamboo: Structure and Culture,* s.l.: University of Duisburg-Essen.

Chapter 6

6 CASE STUDIES OF BAMBOO CONSTRUCTION

6.1 Introduction

Bamboo in its natural form is not commonly used as a structural material in South Africa, being restricted mainly to fencing or used for decorative purposes. As a result, suitable case studies from South Africa were not found, and the case studies presented in this chapter were selected from relevant global examples to illustrate the viability and possibilities available for the use of full-culm bamboo in buildings and structures.

The first two case studies illustrate the use of bamboo construction in scenarios where the common choice of materials would be structural steelwork. The first describes the design and construction of a 52 m span bamboo bridge over a river. In a similar vein, the second case study describes the use of bamboo trusses in a storage warehouse in Bali. Both scenarios illustrate the importance of structural and materials engineering in bamboo construction, where traditional or vernacular construction would be insufficient.

The third case study represents the use of bamboo construction in a residential scenario. It describes a house built entirely from bamboo, which was an entry in the Solar Decathlon Europe, held in Spain in 2010. Although the focus of the Decathlon was the production of self-sufficient buildings regarding energy use, it also clearly demonstrates the use of bamboo for all structural components of a single storey building.

The fourth case study represents the use of bamboo construction in the commercial sector. It expands on the use of bamboo for all building components, describing a community centre in Vietnam which uses local materials and traditional bamboo construction skills to create a modern architectural concept.

6.2 Case Study 1: Bamboo Arched Truss Bridge, Colombia

6.2.1 Introduction

Bridges can be classified in several ways, such as by traffic or material choice. When classifying bridges by structural system, there are six main types, as shown in Figure 6.1. Each of these has specific characteristics, which make them suitable for different applications. Pedestrian bridges often have shorter spans than vehicle bridges, and so the beam or truss types are often chosen for these bridges.

The most common type of truss for a timber bridge with a span of less than 30m is the Pratt or Howe Truss, or a variation thereof (refer Figure 6.2 and Figure 6.3).

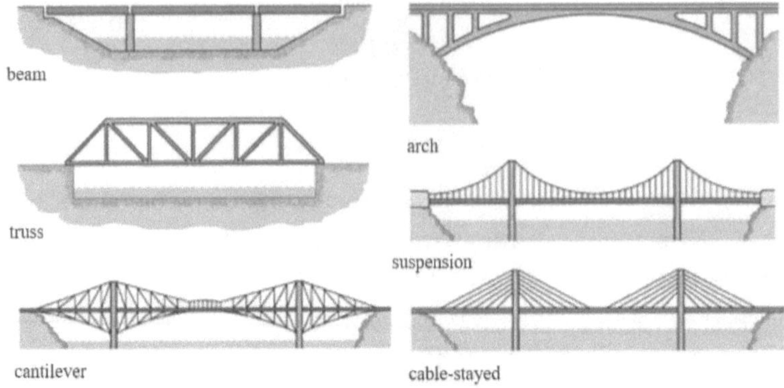

Figure 6.1: Main types of bridges (Billington, et al., 2019)

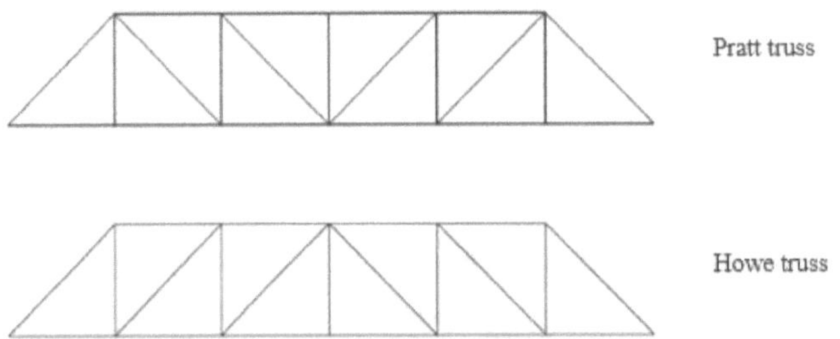

Figure 6.2: Diagram of Pratt and Howe Truss types

Figure 6.3: Kicking Horse Timber Pedestrian Bridge, Canada (Fong, 2006)

For such bridges, the ratio between the truss height and the span length is recommended as 1:10 for timber and bamboo (Stamm, 2009). For longer spans, the height of the truss becomes unwieldy, and arches are used in combination with the trusses, to minimise the height of the side panels while maintaining the recommended arch height (refer Figure 6.4).

Stamm (2009) applied the concept of truss bridges to the use of bamboo culms. In addition, he exploited the natural curvature of the bamboo culms to form a slight upward curvature in the trusses, which enabled the height of the trusses to be minimised. He stated further that the natural curvature of bamboo culms is approximately 2.5% of the length, and thus the curvature of the truss should be similar.

Figure 6.4: Bamboo bridge, Pereira (Stamm, 2009)

However, the addition of a curve to the truss, via an arch or curvature of the material, creates horizontal forces. A conventional flat truss can be designed similar to a simply supported beam, with only vertical reaction forces (refer Figure 6.5 and Figure 6.6). The top and bottom chords behave similarly to the flanges in a girder or I-beam, while the diagonal and vertical members behave in a similar fashion to the web. In comparison, an arch bridge develops axial thrust forces at the supports due to its curvature (refer Figure 6.7), which are then resolved into horizontal and vertical reaction forces

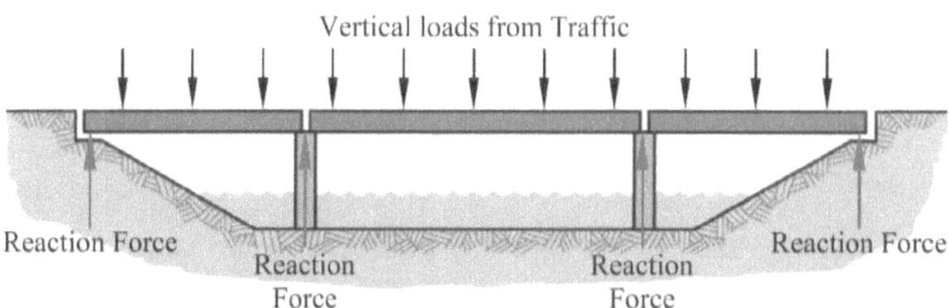

Figure 6.5: Reaction forces for a simply supported beam bridge

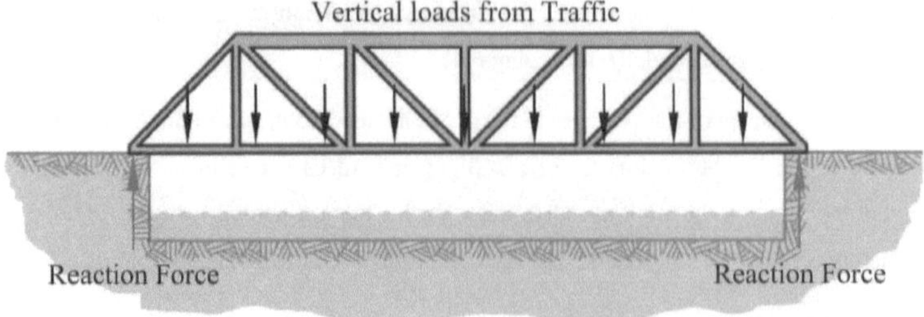

Figure 6.6: Reaction forces for a truss bridge

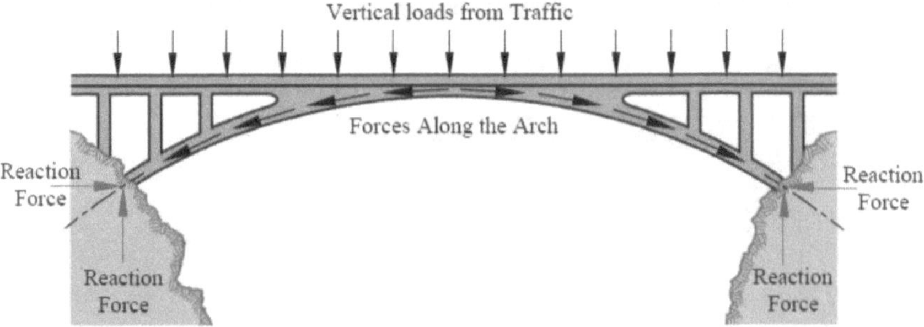

Figure 6.7: Reaction forces for an arch bridge

In an arch bridge, the higher the arch, the smaller the horizontal reactions become at the supports. Conversely, as the arch becomes flatter, so the horizontal reactions would increase and the foundations would need to be designed and detailed accordingly to accommodate this larger force.

Although roofed bridges are uncommon with steel and concrete structures, for timber and bamboo structures, the additional weight of the roof alters the inertia and natural frequencies of the bridge, which assists in resisting the oscillations and vibrations caused by the application of the live loads, either pedestrians or vehicles. The roof also provides the bamboo structure and its users with a measure of weather protection.

6.2.2 Background

In June 1994, an earthquake in the vicinity of Nevado del Huila, a volcano in Colombia, caused a mudslide of approximately 20 m in height to flow down the valley of the Paez River. Approximately 2 000 people were killed, and 120 000 buildings and infrastructure were destroyed (Stamm, 2004).

A relief organisation was set up to reinstate the infrastructure in as short a time as possible. One of the structures to be rebuilt was an equestrian and pedestrian bridge. Stamm proposed a roofed bamboo bridge, using the local species of bamboo, *Guadua angustifolia*, with an engineering basis to avoid collapse in future similar situations. With the assistance of the Rheinisch-Westfälische Technische Hochschule (RWTH) Aachen University in Germany, an arched truss bridge spanning 20 m was designed and built (refer Figure 6.8). The two end spans were designed as cantilevers, with the central piers acting as the only supports. The clear width of the roadway of the bridge was 2 m wide, made from reinforced concrete, and was designed to carry a 2-ton vehicle. The maximum axial force per culm was calculated as 20 kN (Stamm, 2001).

Figure 6.8: Bamboo arch bridge (Stamm, 2004)

Subsequently, Stamm built further bamboo bridges following the same design principles, prior to the Liceo Frances Bridge in Pereira, Colombia. Some of these are listed briefly below.

In 1998, Stamm built a bamboo bridge in Santa Fe de Antioquia, Colombia, with a span of 30 m in approximately one month (refer Figure 6.9). The cost of this bridge was reported as being 30% lower than a comparable steel bridge, as well as having a lower ecological impact (Stamm, 2009).

Figure 6.9: Bamboo Bridge in Santa Fe de Antioquia, Colombia (Stamm, 2009)

In August 2000, Stamm held a Bamboo Bridge Building Seminar together with the German Society for Technical Co-operation (GTZ) and the Technical University of Pereira (UTP). The seminar attendants designed and constructed a bamboo pedestrian bridge with a span of 40 m, crossing four lanes of traffic below, with a 5 m clearance to the walkway (refer Figure 6.10).

Figure 6.10: Bamboo Bridge at the Technical University of Pereira (Stamm, 2009)

The bridge was loaded to 2.5 kN/m² (250 kg/m²), and a vertical deflection of 40 mm was recorded at the centre of the bridge, with an upward recovery of 38 mm on removal of the load. The AASHTO (American Association of State Highway and Transportation Officials) Bridge design specifications state that the vertical deflection for a pedestrian bridge should be less than span/1000. Thus, the initial deflection of 40 mm is within acceptable design limits. The permanent vertical deformation of 2 mm can be ascribed to the "squeezing" or compression of the culm fibres at the contact zones, and the settling of the bolts within the connections. The deflection values, including the recovery value, were considered to be within acceptable safety limits for this type of structure (Stamm, 2009).

Using similar construction plans, a bamboo bridge with a span of 30 m was built over a river in Tierradentro, Inza (refer Figure 6.11).

Figure 6.11: Bamboo Bridge in Tierradentro, Inza (Anon., n.d.)

6.2.3 Liceo Frances Bamboo Bridge

Figure 6.12: Liceo Frances bridge, Colombia (Stamm, 2004)

The Liceo Frances Bridge in Pereira, Colombia, was built in 2001, with a span of 52 m (refer Figure 6.12). This bridge was intended for pedestrian use only. At the time, it cost USD 1000 per meter, which would have been similar to the cost of a similar steel bridge. However, Stamm stated that this span of bridge was reaching the economical limits for bamboo arch bridges (Stamm, 2009).

6.2.4 Construction process

Stamm (2009) reported that a typical *Guadua angustifolia* culm had approximately 200 MPa of tensile strength and 70 MPa of compressive strength. These strengths are is substantially higher than the results reported by Yu (2007) and detailed in Chapter 4 (140 MPa tensile strength, 56 MPa compressive strength), and can be ascribed to the location of the bamboo culms used, with the values based on testing of the local species.

Once the culms had been cut and dried, they were treated with a boric preservative solution.

The side trusses were formed from two compression curves, made from 12 culms bundled together (Stamm, 2004). The bridge had a walkway width of 2.8 m.

The side trusses of the bridge were marked out and assembled on the ground, to ensure the attainment of the correct curvature (refer Figure 6.13).

After assembly, the trusses were lifted into place using cranes, similar to the process depicted for the Pereira bridge (refer Figure 6.14 and Figure 6.15).

Figure 6.13: Typical construction of bamboo trusses for the Liceo Frances Arch Bridge (Stamm, 2004)

As the bridge span was longer than the available culm lengths, it was necessary to connect the culm lengths in such a way as to create continuous members. These connections were made using steel rods inserted in the ends, and the end cavities filled with mortar (Stamm, 2001).

Figure 6.14: Lifting side truss of Pereira bridge (Stamm, 2009)

Figure 6.15: Side trusses in place for Pereira bridge (Stamm, 2009)

The completed bridge is shown in Figure 6.16, indicating the side trusses formed with the 12-culm compression members. The construction details of the bridge were drawn in AutoCAD (refer Figure 6.17 and Figure 6.18), reproduced with translations, courtesy of Stamm (Stamm, 2016).

Figure 6.16: Internal view of the completed bridge (Stamm, 2004)

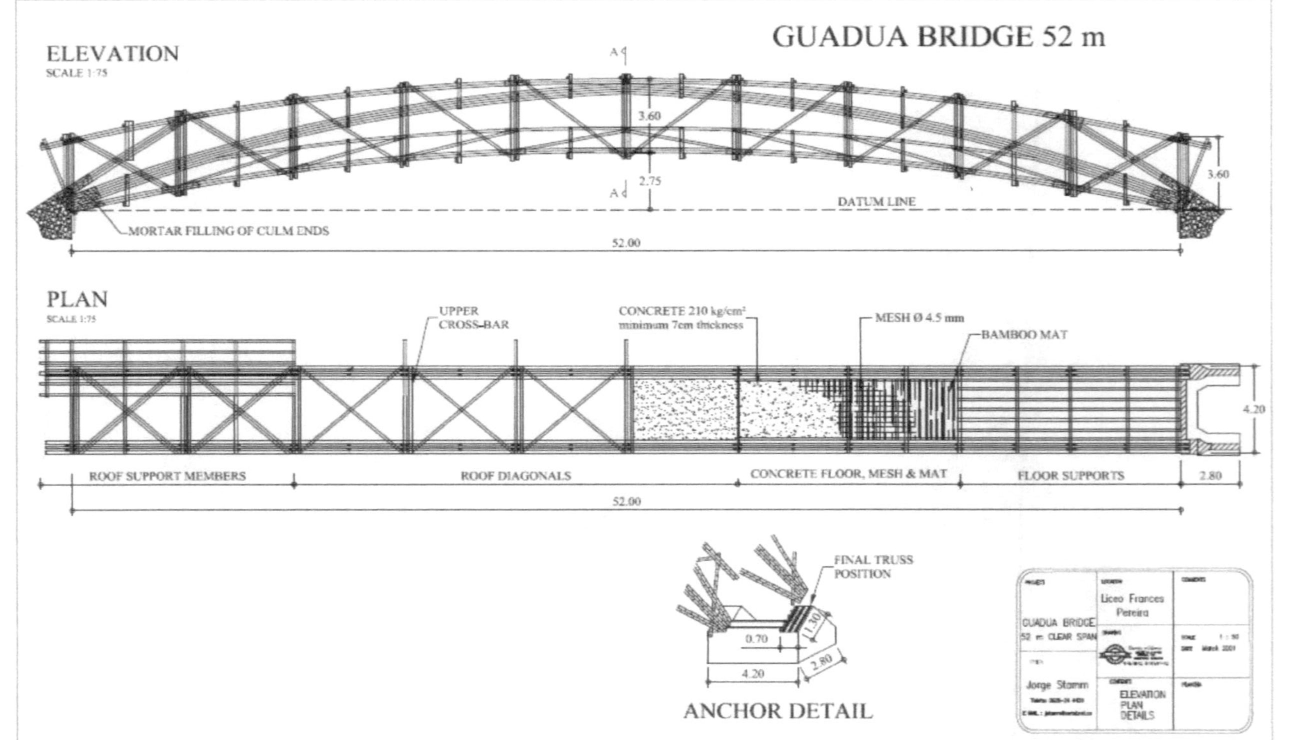

Figure 6.17: Extracts from construction drawing for Liceo Frances Bamboo Bridge, Pereira (translated into English) (Stamm, 2016)

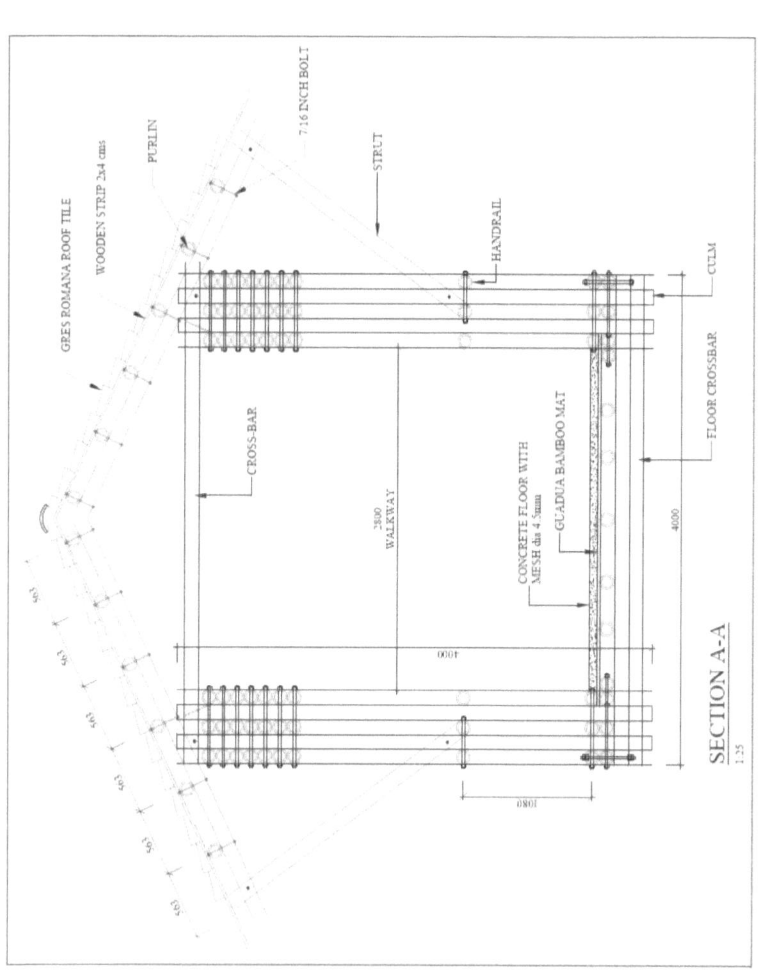

Figure 6.18: Extracts from construction drawing for Liceo Frances Bamboo Bridge, Pereira (translated into English) (Stamm, 2016)

6.2.5   Problems encountered

As a result of various issues as noted below, the construction of the bridge resulted in additional horizontal forces on the abutments, causing approximately 5 cm horizontal movement to the footings. Although the structural engineers deemed this to be within acceptable limits, this horizontal movement affected the performance of the bridge (Stamm, 2016).

After two months, excessive vertical deformation of the trusses was noted. As a result of further analysis, two steel frames were added to stabilise the bridge and limit further vertical movement. The repaired bridge is still in use today (Stamm, 2004).

Five major reasons were given by Stamm for the failure of the bridge (Stamm, 2016).

1)   The first reason given was the height of the arch. A wooden truss would typically require a camber of 1/10 of the span. Although the bamboo trusses were designed on this principle, it was found that a bamboo truss requires an additional camber to counter the deformations. Due to its flexible nature, when this bridge was loaded, the arches lowered by 400 mm. This lowering resulted in compressive forces which the culms were not designed to withstand (refer Figure 6.19).

Figure 6.19: Insufficient arching of the bridge (Stamm, 2004)

2)   A second reason for the poor performance was the lack of classification of the bamboo culms used. The design was based on the structural properties of bamboo culms of the same species but from another region. It was later found that the strength properties were less than assumed, thus

compromising the engineering design. In addition, the culms were not dry when construction started and were harvested from immature plants. When force was applied to the immature culms, cracks formed, and the culms split (refer Figure 6.20). No information was given as to which bridge members were affected, or which forces caused the cracks.

Figure 6.20: Cracked immature culm due to applied force (Stamm, 2004)

3) A third reason for the failure of the bridge was a lack of site control during the construction of the connections. The bridge connections were designed to be filled with a mortar grout, increasing their cross-sectional area and thus also their load-carrying capacity. The bridge itself was designed based on these assumptions. However, a number of these were not done, and so the connections did not perform as expected. The deformations only appeared approximately two months after construction, when it was too late to amend the poor construction or alter the design.

4) A fourth reason was the on-site change in the foundation design. The original design entailed a mass concrete abutment (refer Figure 6.17, Anchor detail). However, the local engineers on the project suggested piling instead. The piles deflected by 50 mm, which is within the design tolerance for piling. The design of the bridge was not adjusted to accommodate this variation, and as a result, the movement of the piles caused unnecessary stress on the culms.

5) A final reason given was that the review engineer on the project amended the number of culms in the main arch of the truss from 15 to 12 culms. This 20% reduction in the culm area would also have reduced the compressive capacity of the arch. Adding this to the reduced structural properties already mentioned, this would imply that the construction would not meet the expected design

loads. This was shown in the excessive deformation and cracking of the culms.

6.2.6 Discussion

This case study demonstrates the extent to which a natural material can be used for a large structure when accompanied by applicable engineering design. However, it also highlights the difficulties encountered with design and construction using a natural material.

Although the structural design might have been adequate, there appears to have been little continuity between the design assumptions concerning the choice of material, and the final selection of the culms to be used. During construction, the designs were changed without the cognisance of the original engineer, with the result that the design of the foundations was no longer adequate, and the deflection criteria were no longer met. In addition, poor quality control on site regarding the application of the design details resulted in inadequate connections and resultant cracks in the members.

Not only is the structural design of such structures important, but as with most structures, it is also necessary to ensure adequate implementation of the design intent and assumptions during construction.

6.3 Case Study 2: Bamboo Warehouse, Bali

Figure 6.21: Bamboo warehouse, Bali (Cayot, 2016)

6.3.1 Background

Asali Bali engineers were approached by a company called PT Tripper Nature in Gianyar, Bali, to design and construct a warehouse from bamboo for their offices in Bali (refer Figure 6.21 and Figure 6.22). PT Tripper Nature sources, process and exports sweet spices from Indonesia.

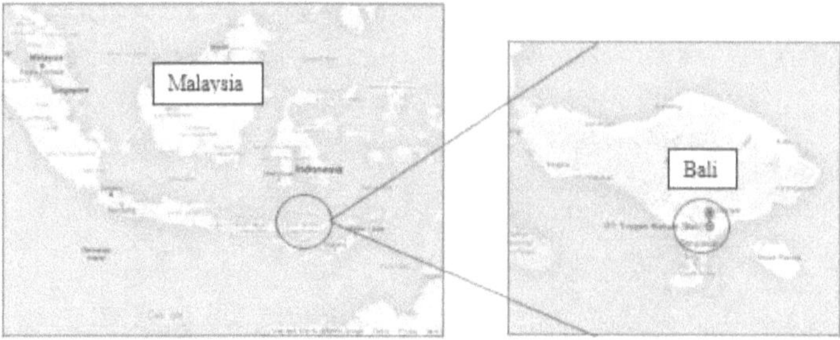

Figure 6.22: Location of site (Google Maps, 2016)

6.3.2 Design and construction processes followed

The specifications for this project included optimisation of the materials and workforce, organic and environmentally friendly treatment of bamboo, and design for easy transport by truck or container.

The prototype warehouse was designed and detailed by D.G.S. Engineers in Singapore. A prototype was built, and seven load tests were conducted to destruction in 2015, to validate the design, although it is unclear as to the exact method or loads used. Although the actual warehouse for PT Tripper Nature measured 33 m long by 16 m wide by 7 m high, the prototype design was for spans of up to 20 m wide, indicating that the prototype design could be used without design modifications (Cayot, 2016).

Using the prototype design, the bamboo culms used were selected from the species *Dendrocalamus asper* and *Gigantochloa Apus*, chosen to all be approximately 140 mm in diameter and carefully selected for straightness. They were then preserved using a product known as Freemite, to prevent attack by insects or fungi. The construction took two months using eight senior artisans who had been trained in bamboo construction.

The trusses were constructed in workshops and transported to site for erection (refer Figure 6.23).

Figure 6.23: Trusses stored on site (Cayot, 2016)

The foundations for the truss columns were designed and detailed in accordance with the Singapore Code of construction (refer Figure 6.24). No details were given regarding the size of the foundations.

Figure 6.24: Truss column foundations (Cayot, 2016)

Once the foundations were cast, the trusses were erected using a crane (refer Figure 6.25 and Figure 6.26).

The roof was clad in steel sheeting, while the sides were covered with thatch, to blend in with the surroundings (refer Figure 6.27 and Figure 6.28).

Figure 6.25: Truss erected on site using a 22 m crane (Cayot, 2016)

Figure 6.26: Final erection of trusses (Cayot, 2016)

Figure 6.27: Completed warehouse (Cayot, 2016)

Figure 6.28: Completed warehouse (Cayot, 2016)

6.3.3   Discussion

Although no engineering design details were available, from inspection of the photographs, it would appear that the trusses are approximately 1.5 m deep. For a structural steel truss for a similar span, the truss depth would be approximately 1 m deep for a lightweight roof. There are diagonal cables visible on the trusses (refer Figure 6.27); thus, these would be considered Pratt trusses (refer Figure 6.2). The columns can be seen to be monopitch trusses, and combine with the roof trusses to form a truss portal system (refer Figure 6.25). The corner diagonals added at the junction of the roof and column trusses obviate the need for knee braces, as this diagonal would provide sufficient stiffness.

Taking the light weight and comparative strength of bamboo culms into consideration, when compared with similar steel sections, this would appear to be a viable alternative to conventional steelwork portal systems.

6.4   Case Study 3: Bamboo Solar House, Spain

6.4.1   Background

Students from Tongji University in China designed and built a bamboo house for the Solar Decathlon Europe 2010, held in Madrid, Spain (refer Figure 6.29). The house has also been named "Sunshine Inn" due to its solar panels that generate electricity to power the house. The house is said to have minimised carbon dioxide emissions throughout the entire production phase (Anon., 2010), although exact details and values were not available.

Figure 6.29: Exterior view of the completed house (Anon., 2010)

6.4.2   Design and construction processes followed

The bamboo used to construct the house was sourced locally in China. Although no mention is made of the exact species, the most common species found in China is *Phyllostachys pubescens*.

The 75 m² bamboo house had two curved and sloping roofs and was almost entirely constructed from bamboo. One wall was clad with photovoltaic panels, which generated sufficient electricity to power the one-bedroom house. The interior walls were covered with phase change materials, similar to those used in heat packs, to heat and cool the house. From the photographs, it would appear that the house also had insulation of some sort on the walls (refer Figure 6.33).

Tongji University's team consisted of 20 members, composed of doctoral, postgraduate and undergraduate students ranging across many disciplines, such as architecture, urban planning and energy development, although no mention was made of any structural engineering input. The team spent six months designing and constructing the house. In order to construct the house rapidly on site, the team prefabricated most of the house at the University before shipping to Madrid for the Decathlon.

The house was designed as a framed structure based on traditional Chinese architecture (refer Figure 6.30 to Figure 6.33), with smaller bamboo culms to form the infill wall panels. The concept was aimed at combining modern architecture and energy supply with a traditional Chinese building method and material.

Figure 6.30: Construction and preparation of bamboo culms at Tongji University (Anon., 2010)

Figure 6.31: Construction of the house at Tongji University (Anon., 2010)

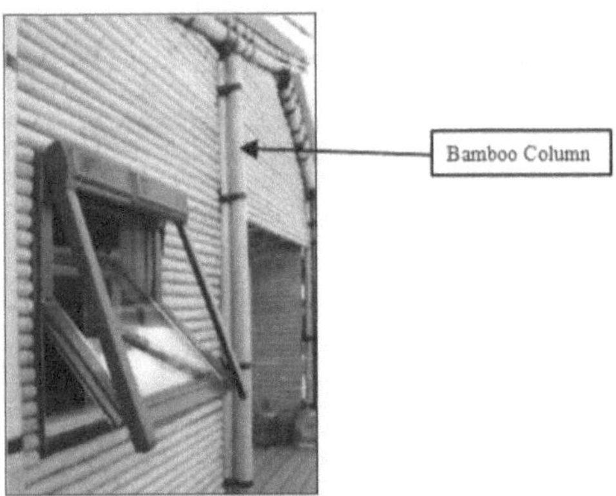

Figure 6.32: Window and column detail (Anon., 2010)

Figure 6.33: Construction phase on site (Anon., 2010)

6.4.3 Discussion

Although it has been mentioned that it took six months to construct this house, it is unclear as to how much of this time was spent on design, and how much was actual construction. Other than the photographs, no design information of the house was available, so it is unclear whether the house was built from traditional methods, or whether any structural design was incorporated. The photographs show steel bracing cables and straps acting as connectors for the culms. This would seem to indicate that modern structural techniques were applied to a traditional material and construction method. However, there was no mention of any design codes used, and no record was found of what happened to the house after the Decathlon.

Although this case study presented an example of what could be done with bamboo culms in a residential structure, from an engineering perspective, it is inconclusive as to whether this is viable or not.

6.5 Case Study 4: Diamond Island Community Centre, Vietnam

Figure 6.34: Diamond Island Community Centre (Nghia, 2015)

6.5.1 Background

Diamond Island is a group of high-rise condominiums on an artificial island on the Saigon River, adjacent to the city of Ho Chi Minh, Vietnam (refer Figure 6.35). It was developed by the Kusto Group, an international investment company, together with Binh Thien An, a real estate business rebranded to Kusto Vietnam in 2012. The community centre (refer Figure 6.34) was designed as a multi-functional centre for the residents of the condominiums, as well as for visiting tourists. It was designed by Vo Trong Nghia Architects and built by Wind and Water House JSC (Joint Stock Company) contractors and completed in August 2015. It was nominated as the Building of the Year 2016 by Arch Daily, an online architectural website.

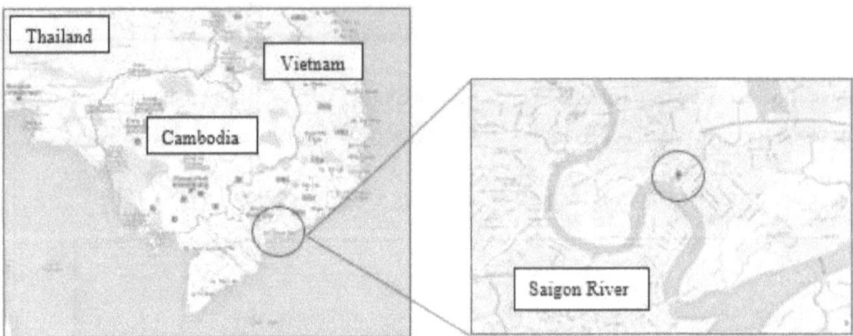

Figure 6.35: Location of Diamond Island (Google Maps, 2016)

6.5.2 Design and construction processes followed

Bamboo grows abundantly, quickly and cheaply in Vietnam, where culms cost approximately USD 1 in 2014. Bamboo is traditionally used to create baskets, tableware and furniture, and this inspired the choice of bamboo as a structural material for the pavilions. The architect, Vo Trong Nghia, is quoted as saying that bamboo building is suitable to climates such as Vietnam, where little or no winter exists.

In Vietnam, there are two main types of bamboo, namely Tam Vong (*Dendrocalamus strictus*) and Tre Gai (*Bambusa stenostachya*). Tam Vong, also known as "Iron Bamboo" or "Calcutta Bamboo" has a relatively small diameter of up to 80 mm and is often chosen for its flexibility. Tre Gai has been approved by ICE for use as structural poles (refer Chapter 4 for more details). Although no detail was given, it is assumed that these two species were used in the construction of the community centre, due to their abundance in the area.

The community centre consists of two large and six small bamboo pavilions, used for events such as parties and conferences (refer Figure 6.34). There are plans to convert the pavilions to restaurants at a later stage.

The two large pavilions are in the form of domes. They are 24 m in diameter, and 12.5 m in height. The architect stated that the bamboo configuration was inspired by the bamboo baskets used for poultry by the local Vietnamese residents. The diagonal bamboo members were constructed and woven on site by skilled workers. All connections were constructed using bamboo pegs and rope, with no metal joints (refer Figure 6.36).

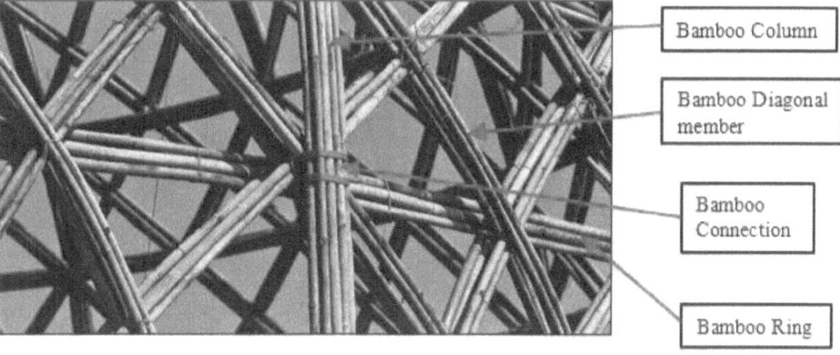

Figure 6.36: Typical connection of column and ring (Nghia, 2015)

Figure 6.37: Large dome pavilion (Nghia, 2015)

They consist of a double-layered dome structure (refer Figure 6.38). The outer layer of thatch overhangs the inner bamboo structure, creating a deep eaves section and protecting the bamboo from sun and rain.

Beneath the thatch roof, lengths of bamboo are bound together to form columns, which in turn extend up to bamboo rings which form the dome-shaped roof (refer Figure 6.39, Figure 6.40 and Figure 6.41).

All the domes have a skylight or opening at the peak of the roof, which allows hot air to escape upwards.

The skylight together with the open periphery provides sufficient light during the day to obviate the need for artificial lighting (refer Figure 6.42).

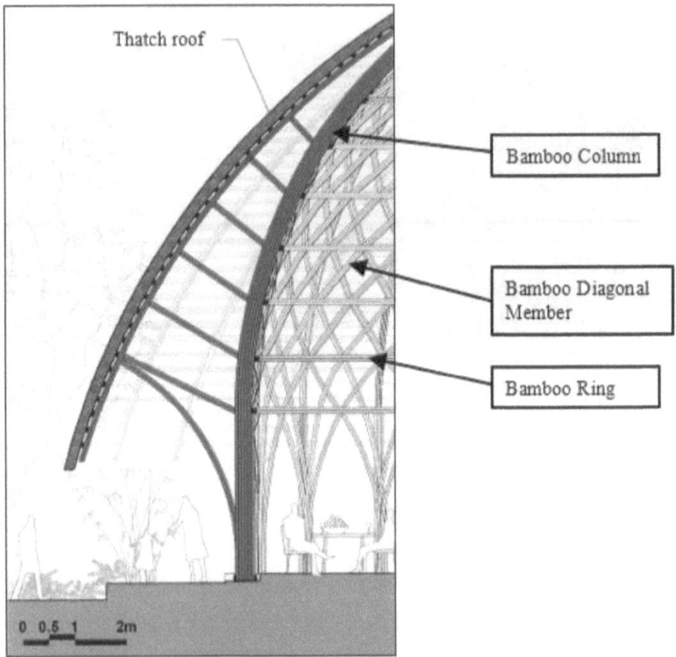

Figure 6.38: Typical cross-section of larger dome structure (Nghia, 2015)

Figure 6.39: Pavilions during construction (Nghia, 2015)

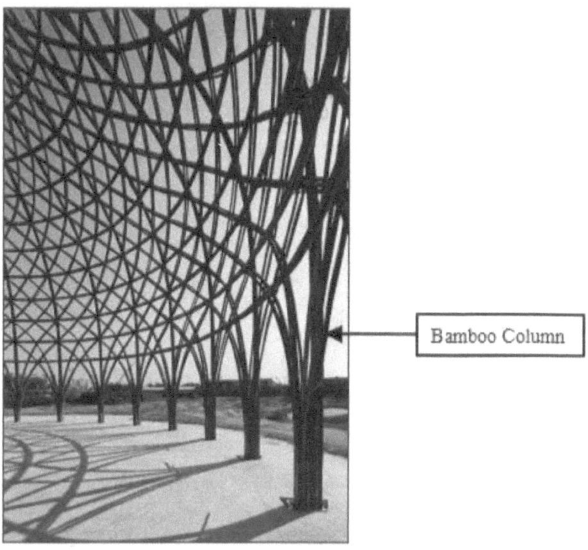

Figure 6.40: Interior view of a pavilion during construction (Nghia, 2015)

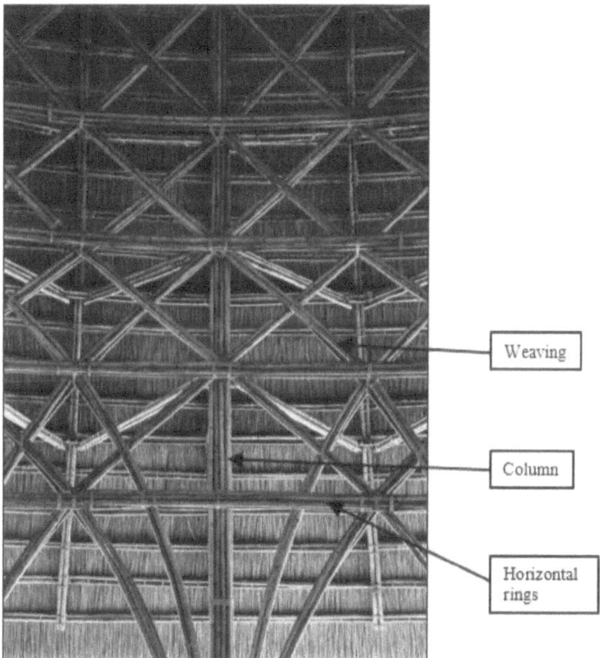

Figure 6.41: Internal column connections (Nghia, 2015)

Figure 6.42: Skylight in a pavilion (Nghia, 2015)

The six smaller pavilions are shaped like umbrellas (refer Figure 6.43 and Figure 6.44). These are 11 m in diameter, and 7 m in height. The units were prefabricated off-site in 12 units, and assembled on site, to improve the efficiency and speed of construction.

Figure 6.43: Comparison of larger and smaller pavilions (Nghia, 2015)

Figure 6.44: Smaller pavilion (Nghia, 2015)

6.5.3 Discussion

This case study illustrates the use of a local resource in creating an aesthetic and functional structure.

Although the species of bamboo used has a large diameter (150 mm), the columns are formed by connecting several culms. There do not seem to be connectors between the culms, so there is doubt as to whether the full capacity of the multiple-culm column was used. In addition, the culms needed to be joined to create the desired lengths. This joining was achieved by means of a form of strapping. An alternative connection detail could reduce the bulkiness of the column connections.

Lateral stability to the domes was provided by the bamboo rings, which would act in hoop tension, while overall stability was provided by connecting the ring of columns. Although the base connection details are not visible in the images, they could be expected to have some moment capacity.

Although no information was found regarding structural engineering input, the architectural firm Vo Trong Nghia has in-house engineers and reports that their architects and engineers work closely on projects.

6.6 Conclusions

These case studies demonstrate that bamboo culms are suitable for use in large structures, such as bridges or trusses, as well as for smaller structures such as buildings or houses. The bamboo species used in the bridge case study was *Guadua angustifolia*, with a reported compressive strength of 70 MPa and tensile strength of 200 MPa. The warehouse bamboo was a combination of two species, namely *Dendrocalamus asper* and *Gigantochloa apus*. Although no strengths were given for the actual bamboo used, the values reported for *Dendrocalamus asper* in Chapter 4 are 31 – 64 MPa compressive strength and 137 - 222 MPa for the tensile strength, and 34 MPa compressive strength, 178 – 299 tensile strength for *Gigantochloa apus*.

Although no mention was made of the bamboo species used for the Solar House case study, it was assumed that this was constructed from *Phyllostachys pubescens*, one of the most common species in China. The compressive and tensile strengths as reported in Chapter 4 for this species are 56 MPa and 196 MPa, respectively.

Similarly, for the Diamond Island Community Centre, it was assumed that this was constructed from the abundant local species, namely *Bambusa stenostachya*, with a compressive strength of 51 - 66 MPa and tensile strength of 8 MPa.

Table 6.1: Compression and Tensile strengths of Bamboo species (extracted from Chapter 4)

Bamboo species	Compression strength parallel to the grain (MPa)	Tensile strength parallel to the grain (MPa)
Bambusa stenostachya	51 - 66	8
Dendrocalamus asper	31 - 64	137 - 222
Gigantochloa apus	34	178 - 299
Guadua angustifolia	56	140
Phyllostachys pubescens	56	196

Apart from *Bambusa stenostachya*, which appears to have an extremely low tensile strength value, the other bamboo species used in these four case studies have compressive and tensile strengths in similar ranges.

In South Africa, the species considered suitable for construction is *Bambusa balcooa*. From the literature, this species is reported to have a wall thickness of 25 – 50 mm, a compressive strength of 48 - 81 MPa and tensile strength of 107 - 228 MPa. Based on its similarity in compressive and tensile strength to the above-mentioned species, it would appear that this South African species would perform in a similar manner if used in such structures. However, there is as yet no conclusive evidence of this. The following chapter will deal with this in more detail.

6.7 References

Anon., 2010. *Bamboo house at Solar Decathlon.* [Online] Available at: http://www.designboom.com/architecture/bamboo-house-at-solar-decathlon/ [Accessed 4 February 2016].

Anon., n.d. *Joerg Stamm.* [Online] Available at: http://bambus.rwth-aachen.de/eng/reports/joerg_stamm/referatstamm.html [Accessed 26 April 2016].

Billington, P., Shirley-Smith, H. & Billington, D., 2019. *Bridge Engineering.* [Online] Available at: https://www.britannica.com/technology/bridge-engineering [Accessed 21 October 2019].

Cayot, T., 2016. *Asali Bali* [Interview] (24 February 2016).

Cayot, T., 2016. *Designboom.* [Online] Available at: http://www.designboom.com/architecture/asali-bali-engineered-bamboo-structure-bali [Accessed 20 March 2016].

Fong, V., 2006. *Kicking Horse Pedestrian Bridge [image].* [Online] Available at: https://commons.wikimedia.org/wiki/File:Kicking_Horse_Pedestrian_Bridge.jpg

Google Maps, 2016. *Bali [image].* [Online] [Accessed 23 March 2016].

Google Maps, 2016. *Vietnam [image].* [Online] Available at: https://www.google.co.za/maps/place/Vietnam/@15.9030623,105.8066 [Accessed 1 February 2016].

Nghia, V. T., 2015. *Vo Trong Nghia Architects.* [Online] Available at: http://www.votrongnghia.com/projects/diamond-island-community-hall [Accessed 27 January 2016].

Stamm, J., 2001. *Bamboo Bridges as an alternative to rainforest destruction.* Ecuador, INBAR.

Stamm, J., 2004. Analysis of Techniques and Work methods of Traditional European Carpentry applied during 10 Years in Covered Bamboo bridges.

Stamm, J., 2009. *Seven Concepts to Build A Bamboo Bridge.* Thailand, s.n., pp. 70-100.

Stamm, J., 2016. *Email correspondence with Author.* s.l.: s.n.

Chapter 7

7 SUITABILITY OF BAMBOO CONSTRUCTION FOR SOUTH AFRICA

7.1 Introduction

The previous chapters have illustrated that bamboo is suitable for use as a structural building material, both from an engineering perspective as well as from the viewpoint of sustainability. However, the question remains, is bamboo a viable building material in South Africa?

To address this question, the following issues will be addressed in this chapter:

1) Is the bamboo that is growing in Southern Africa suitable for construction purposes?
2) Is there sufficient bamboo to satisfy construction needs?
3) What are the engineering and material properties of South African bamboo?
4) What applicable guidelines and codes are available for South African conditions?
5) Does South Africa have the necessary skills for bamboo construction?
6) Are there any examples of bamboo construction in South Africa?
7) Are there limitations to using bamboo in South Africa?

7.2 Bamboo in South Africa and surrounds

Africa has 7% of the world's bamboo resources, with a variety of bamboo species in South Africa and the surrounding countries. Some of the species are *indigenous*, being found in a specific area due to natural processes. Others are considered as *introduced* species, having been brought to an area by human behaviour or intervention. After ten years without direct intervention by humans, introduced species can be considered to be *naturalised*.

7.2.1 Indigenous bamboo species

Africa has approximately one million hectares of indigenous bamboo (Scheba, et al., 2017). In 2011, over 850,000 hectares of lowland bamboo, *Oxytenanthera abyssinica*, and over 130,000 hectares of highland bamboo, *Arundinaria alpina*, were reported in Ethiopia (Kibwage, et al., 2011). Further to this, in 2012, 2.7 million hectares of bamboo forest was reported by six African countries (Ethiopia, Kenya, Nigeria, Uganda, Tanzania and Zimbabwe) (Bamboo Interim Steering Committee, 2012).

Of the native species of bamboo found in Africa, three of these are typically used for construction, namely *Arundinaria alpina*, *Oreobambos buchwaldi* and *Oxytenanthera abyssinica* (Bystriakova, et al., 2002).

Arundinaria alpina, also known as *Yushania alpina* or African alpine bamboo, can be found in Eastern Africa, from Sudan in the north to Zambia in the south, at altitudes of 2200 to 3500 m above sea level

(INBAR, 2015) as indicated in Figure 7.1.

Figure 7.1: African countries reporting the presence of *Arundinaria alpina* (Spears, 2008)

Oreobambus buchwaldi is found in the countries of Zimbabwe, Zambia and Malawi, at altitudes of 300 to 2000 m above sea level (INBAR, 2015) (refer Figure 7.2).

Figure 7.2: African countries reporting the presence of *Oreobambus buchwaldi* (Bystriakova, et al., 2004)

South Africa has only one indigenous species of bamboo, namely *Thamnocalamus tessellatus* or "bergbamboes" (refer Figure 7.3). This species is a loosely tufted bamboo that grows in dense clumps. It is used to construct gates and screens and used as walking sticks, implement handles and spears (Scheba, et al., 2017).

Figure 7.3: *Thamnocalamus tessellatus* **(Lekhetho & Notten, 2019)**

It is found in the high mountains of South Africa, Lesotho and Swaziland, at elevations of 1500 – 2000 m, as indicated in Figure 7.4.

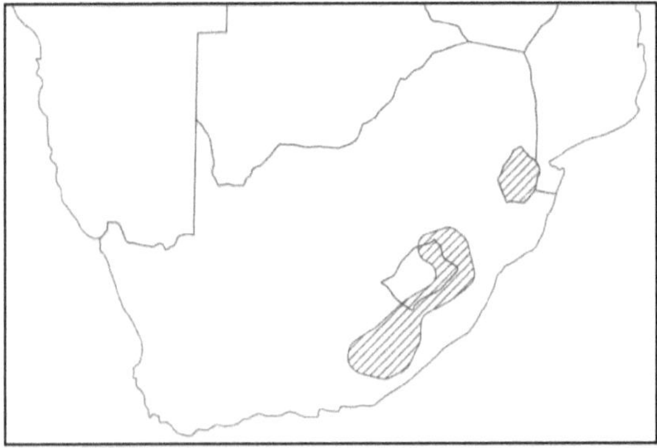

Figure 7.4: Distribution of *Thamnocalamus tessellatus* **in South Africa (Victor, 2006)**

7.2.2 Introduced bamboo species

Numerous species have been introduced into Kenya and Uganda. The most successful of these are *Bambusa bambos, Bambusa tulda, Bambusa vulgaris, Dendrocalamus asper, Dendrocalamus brandisii, Dendrocalamus membranaceus, Dendrocalamus strictus, Phyllostachys pubescens* and *Thyrsostachys siamensis.*

Oxytenanthera abyssinica, or Holy Venda Bamboo, can be found across the width of Africa, from Ethiopia to northern South Africa (refer Figure 7.5). This species is thought to have been introduced to South Africa by the Venda tribe when they migrated from Zimbabwe to the current Limpopo Province of South Africa (Canavan, et al., 2018; Scheba, et al., 2017). It is traditionally used for ceremonial purposes, basketry and furniture, as well as small-scale construction.

Figure 7.5: African countries reporting the presence of *Oxytenanthera abyssinica* (Bystriakova, et al., 2004)

Another introduced bamboo species, now considered naturalised, is *Bambusa Balcooa*, also known as "Giant bamboo", which was purportedly introduced in South Africa by the Dutch East India Company in 1653 (Canavan, et al., 2018). This species is found in the provinces of the Western Cape (refer Figure 7.6 and Figure 7.7), the Eastern Cape and Kwazulu-Natal (van Asch, 2009).

Figure 7.6: *Bambusa balcooa* at Canal Walk Shopping Centre, Western Cape (S. Ross, 2017)

Figure 7.7: ***Bambusa balcooa*** **at Canal Walk Shopping Centre, Western Cape (S. Ross, 2017)**

7.2.3 Commercial Initiatives

Several commercial initiatives have been started in South Africa in the past ten years (refer Figure 7.8).

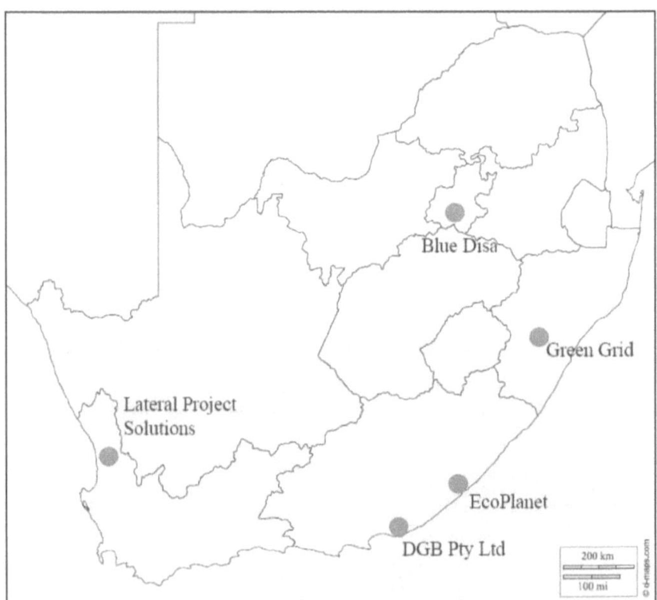

Figure 7.8: Map of commercial bamboo projects and activities in South Africa (adapted from Scheba, et al., 2017)

- *DGB Pty Ltd, Eastern Cape*

A South African winery and distillery company, DGB, started a community-based bamboo project of 10 hectares in size in Mooiplaas, near East London, at the end of 2011 (Scheba, et al., 2017). Unfortunately, many of the initial plants were lost due to fire, weeds and drought, but half the area was replanted in 2015 with additional *Bambusa balcooa* seedlings.

- *EcoPlanet Bamboo South Africa, Bathurst, Eastern Cape*

EcoPlanet Bamboo Southern Africa, which is part of the EcoPlanet Bamboo Group, a worldwide owner of bamboo plantations, has owned and developed the Kowie Bamboo Farm in Bathurst in the Eastern Cape since 2012 (refer Figure 7.9). The 482-hectare farm grows two species of bamboo, namely *Bambusa balcooa* and *Oxytenanthera abyssinica*, which are manufactured on-site into carbon products for water and air purification (Anon., 2017; Scheba, et al., 2017).

Figure 7.9: Kowie Bamboo Farm, Bathurst, Eastern Cape (Anon., 2017)

- *Green Grid Beema Bamboo Project, Kwazulu-Natal*

The Green Grid Beema Bamboo project started in the Ilembe District in Kwazulu-Natal in November 2013, with the planting of the first bamboo shoots in April 2014. Beema Bamboo (also spelt *Bheema Bamboo*) is a bamboo clone developed by Dr Barathi from the species *Bambusa balcooa*. It is sterile, non-invasive clumping variety with a nearly solid culm (refer Figure 7.10) of 80 – 150 mm.

The Green Grid project aimed to cultivate Beema Bamboo as biomass feedstock to generate 'green electricity' for the SAPPI (South African Pulp and Paper Industries) Tugela Mill using gasification. Despite delays due to drought, land ownership disputes, community conflicts and late receipt of the bamboo shoots, electricity generation was expected to commence in 2017. Unfortunately, due to a shortage of funds, the power generation project was abandoned (N Barathi 2020, personal correspondence, 20 October).

Figure 7.10: Cross-section of *Beema Bamboo* (Barathi, 2019)

- *Blue Disa Co-operative, Gauteng*

Blue Disa is an agricultural co-operative located in Lawley, south-west of Johannesburg in Gauteng Province. Through the organisation Food and Trees for Africa's *Bamboo for Africa* programme, 4600 *Bambusa balcooa* seedlings were received and planted in February 2014. These bamboo plants are intended to form part of a carbon sequestration project (Scheba, et al., 2017).

- *ECDC Projects, Eastern Cape*

In August 2011, a symposium was held in East London, Eastern Cape, South Africa, by the Industrial Development Corporation (IDC) and the Eastern Cape Development Corporation (ECDC), to promote bamboo development in South Africa. One of the outcomes of the symposium was the start of pilot projects to cultivate bamboo. The first consisted of a plot one hectare in size at St Albans near Port Elizabeth, with two larger projects of five hectares each in Centani in the former Transkei and at Ndakana near Stutterheim (van Rijswijck, 2011), with an investment of R1 million. A further R2 million was invested in expanding the Ndakana project to 100 hectares. These sites were intended to grow into large bamboo forests, covering a minimum of 500 hectares (Scheba, et al., 2017). Unfortunately, these projects were discontinued due to lack of funding and the death of the plants at the sites (Scheba, et al., 2017).

A further outcome of the symposium was the formation of an interim Bamboo Steering Committee, with the mandate to form a nationally representative industrial body. The Committee produced a strategy document in 2012, "Business Case for the National Bamboo Association of South Africa NBASA", proposing the institutional framework and functions (Scheba, et al., 2017).

- *Lateral Project Solutions, Vredendal*

A farmer by the name of Joubert Roux is growing approximately 60 different species of bamboo on a plot of 1 hectare outside Vredendal in the Western Cape (refer Figure 7.11 and Figure 7.12). He has imported various species, such as *Phyllostachys edulis*, *Bambusa vulgaris* and *Oxytenanthera abyssinica*. He is growing them to determine which ones adapt best to the local South African climate, as found in Vredendal (classified as a temperate coastal climate, according to SANS 10400-XA (SANS 10400-XA, 2011). From this, suitable species for construction or other end-purposes can be selected. He has not yet published his findings; thus, it is unclear as to whether species other than the ones documented as suitable for construction, should be considered or not.

Figure 7.11: Lateral Project Solutions bamboo farm, Vredendal (Canavan, et al., 2018)

Figure 7.12: *Oxytenanthera abyssinica* growing at Lateral Project Solutions bamboo farm, Vredendal (NBASA, 2017)

7.3 Quantitative requirements for bamboo construction

South Africa currently imports a large number of its bamboo products. In 2012, 68 tons of raw bamboo were imported, increasing to 100 tons in 2016 and 128 tons in 2018 (INBAR, 2020). This increase in imports would indicate an increased demand for bamboo in South Africa (Scheba, et al., 2017), although this is small in comparison to world imports such as China, which imported 11 650 tons of raw bamboo in 2018. The most common uses of bamboo were for fencing, flooring, cladding, home products such as chopping boards and household furniture.

South Africa was also the fourth highest exporter of bamboo flooring and the seventh-highest exporter of bamboo paper-based articles in 2017 (INBAR, 2019) (refer Table 7.1 and Table 7.2).

Table 7.1: Main traders of bamboo flooring in 2017 (INBAR, 2000)

Rank	Exporters		Importers	
	Country	Value (USD thousand)	Country	Value (USD thousand)
1	China	212 701	EU	25 347
2	EU	5 707	Australia	9 827
3	Ghana	3 507	Malaysia	8 674
4	**South Africa**	**1 571**	New Zealand	4 044
5	USA	1 452	USA	3 883
6	Serbia	998	Israel	2 636
7	United Arab Emirates	365	Botswana	2 142
8	India	352	South Korea	2 052
9	Australia	223	Singapore	1 674
10	Malaysia	198	Sri Lanka	1 480
	World	**227 585**	**World**	**80 895**

Table 7.2: Main traders of bamboo paper-based products (INBAR, 2019)

Rank	Exporters		Importers	
	Country	Value (USD thousand)	Country	Value (USD thousand)
1	China	12 035	EU	22 901
2	EU	8 868	USA	4 167

Table 7.2 (continued):

Rank	Exporters		Importers	
	Country	Value (USD thousand)	Country	Value (USD thousand)
3	USA	758	Singapore	1 159
4	United Arab Emirates	143	Iran	1 088
5	Singapore	141	Saudi Arabia	609
6	Canada	134	Canada	573
7	**South Africa**	**130**	Australia	417
8	Russia	85	Bangladesh	374
9	Thailand	64	Dominica	301
10	Dominica	26	New Zealand	294
	World	**23 127**	**World**	**35 364**

In 2012, the Bamboo Interim Steering Committee produced a strategy document for the promotion of bamboo in South Africa. Included in this document was a desktop study done by the South African Agricultural Research Council (ARC) regarding the potential production areas for bamboo in South Africa. They reported that South Africa had 421 715 hectares with a high potential for bamboo production, representing 0.5% of the total RSA surface area. Kwazulu-Natal had the largest suitable area (188 023 ha), while the Western Cape had the smallest (6214 ha). It was noted in the report that this represented the potential for bamboo production, and not the available land (Bamboo Interim Steering Committee, 2012).

The summary of these results is found in Table 7.3 on the following page.

Based on the 2012 import information (68 tons of raw bamboo), the Bamboo Steering Committee calculated that 865 ha of bamboo farming would be sufficient to produce that amount of bamboo locally, based on the assumption that the required bamboo varieties can be grown in South Africa. This calculation was based on the estimation by the Eastern Cape Development Corporation (ECDC), which estimated that one hectare of bamboo plantation could yield 20 to 40 tons of bamboo per year. If the demand for bamboo increased even by 550%, such as would be expected if bamboo were used as a construction material, it was estimated that 4748 ha (1% of high potential area) would be sufficient to satisfy total demand. Thus, it would seem that the Steering Committee believed that South Africa could produce sufficient bamboo locally to supply current and future bamboo needs, without the need for importing any bamboo.

Table 7.3: Potential bamboo production areas in South Africa (Bamboo Interim Steering Committee, 2012)

Province	Province area (ha)	Suitability class (ha)				Total (ha)	Total (%)
		High	**Moderate**	**Low**	**Marginal**		
Eastern Cape	16,896,598	42,839	315,078	489,468	465,860	1,313,245	1.5%
Free State	12,982,516	-	26,655	219,964	659,589	906,208	1.1%
Gauteng	1,817,830	-	-	2,264	666,768	669,032	0.8%
KwaZulu-Natal	9,435,133	188,023	916,221	914,117	684,986	2,703,347	3.1%
Mpumalanga	7,649,471	147,902	271,680	650,061	992,269	2,061,912	2.4%
Limpopo	12,575,396	36,737	49,263	82,463	722,506	890,969	1.0%
North West	10,488,168	-	-	-	364,025	364,025	0.4%
Western Cape	14,010,998	6,214	11,179	16,630	48,468	82,491	0.1%
Total SA	**85,856,110**	**421,715**	**1,590,076**	**2,374,967**	**4,604,471**	**8,991,229**	**10.5%**
% Total RSA surface area		**0.50%**	**1.9%**	**2.8%**	**5.4%**	**10.5%**	

As yet, no figures are available as to the amount of bamboo produced locally by the various growers and nurseries. Once these figures become available, it would be possible to verify the above assumptions regarding the amount of bamboo produced, and thus whether South Africa can meet the bamboo demand with locally grown bamboo.

7.4 Engineering and material properties of Southern African bamboo

Although there are possibilities to use other species of bamboo for construction, the species in South Africa currently considered suitable is *Bambusa balcooa*. This species is also one of the species recommended by INBAR for use as a structural element (Richard, 2013).

Bambusa balcooa is a clumping species, which does not send out underground shoots, and therefore is not considered an invasive species. These clumps have been reported to expand to a diameter of approximately 5 m (INBAR, 2000).

Bamboo plants prefer growing in well-drained soil, with the water requirements depending on the individual species and location. The total water requirement for these plants has been reported as 5000 litres per year per clump (Food & Trees for Africa, 2011). This approximates to 14 litres per day, which is similar to the amount reported by Roux (NBASA, 2017) of 14.5 litres per day during the summer season. As the amount of water provided influences the growth rate, Scheba (2017) recommends that the bamboo be grown in areas of highest rainfall, such as the Eastern Cape and Kwazulu-Natal, with a mean annual rainfall of 600 – 1000 mm per year (refer Figure 7.13). Growing of bamboo species in other areas would likely require additional irrigation to meet the water requirements.

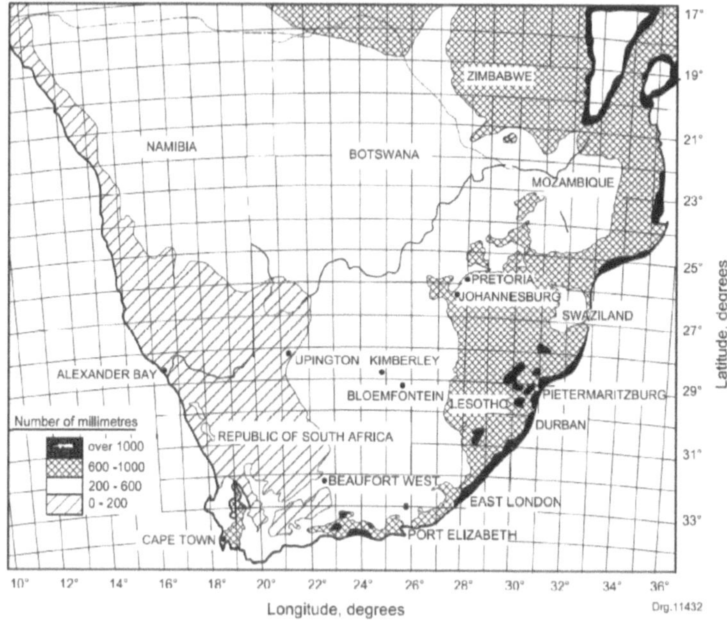

Figure 7.13: Mean annual rainfall in South Africa (SANS 10400-K, 2015)

The engineering and material properties of *Bambusa balcooa* as reported in Chapter 4 are summarised in Table 7.4.

Table 7.4: Summary of the properties of *Bambusa balcooa*

	Density (kg/m³)	Modulus of Rupture (GPa)	Compressive strength (MPa)	Shear strength (MPa)
Bureau of Indian Standards, 2015	783 (green)	65.4 (green)	60.6	n.a.
Naik, 2005	720 - 850	116 - 173	48 - 81	10.1 – 13.1
INBAR, 2000	n.a.	69.6 – 92.8	51 – 57.3	n.a.
Average	**785**	**93**	**60**	**12**

Note: n.a. = not available

No South African data are available regarding the material and engineering properties of this species of bamboo. Therefore, testing would be required of the culms to be used, before the detailed design of any structure. However, the preliminary conceptual design could be based on either the values given above or general rules of thumb. Janssen (2000) provided some ratios, where the ultimate stress in Megapascals (MPa) can be calculated from a ratio multiplied by the density in kg/m³. These ratios and resulting stresses for *Bambusa balcooa* are given in Table 7.5, assuming an average density of 785 kg/m³.

Table 7.5: Ratios and resulting stresses for mechanical properties of bamboo (assuming an average density of 785 kg/m³) (Janssen, 2000)

	Modulus of Rupture	**Compression**	**Shear**
Ratio	0.14	0.094	0.021
Stress (MPa)	110	73.8	16.5

When these stresses are compared with the reported values, it would seem that the ratios give values that are higher than actuality. It is recommended these ratios be used with caution, or that further verification is used before accepting these calculated stress values.

7.5   Guidelines and Codes for Bamboo Construction in South Africa

There are currently no South African design codes for the design or construction of bamboo structures. However, there are codes from other countries that may be used, such as the National Building Code of India or the International Code for Bamboo, ISO 22156 & 22157.

In addition, INBAR has published several guidelines and material values that can be used for the preliminary sizing and design of such structures, such as *Designing and Building with Bamboo – Technical Report 20* (Janssen, 2000) and *Design of Bamboo Scaffolds – Technical Report 23* (Chung & Chan, 2002).

As bamboo construction is not yet an accepted form of building construction in South Africa, as defined by the NHBRC (National Home Builders Registration Council), additional tests and calculations would be required in order to obtain building permits, such as load-bearing capacity and behaviour in fire. Although reference can be made to the ISO standards in a rational design, these calculations would still be required.

Many Eastern countries, such as China and Japan, have a history of bamboo construction and have adapted their construction and design methods to suit the vagaries of their local bamboo species. These adaptations can be transferred in principle to bamboo construction in other countries where bamboo construction is uncommon, such as South Africa.

7.6   Construction skills required for bamboo construction

In order to construct a bamboo building, simple carpentry skills and tools are required. There are numerous examples of bamboo buildings being built in rural areas in countries such as Colombia, where the local people were taught the necessary skills during construction. This transfer of knowledge could be applied in South Africa, where there is a large proportion of unskilled labour.

INBAR has also published several of guidelines for bamboo construction, such as *Design Guide for Engineered Bahareque Housing* (Kaminski, et al., 2016), *Designing and Building with Bamboo*

(Janssen, 2000) and *Bamboo School Building* (Paudel & Ayeh, 2003). Other publications are also available, such as *Construction Manual with Bamboo* (Stamm, et al., 2014).

The United Nations Industrial Development Organization (UNIDO) listed several hand and power tool examples for bamboo construction (refer Figure 7.14). Minimal labour skills are required to operate these, resulting in a high level of constructability (United Nations Industrial Development Organization (UNIDO), 2008).

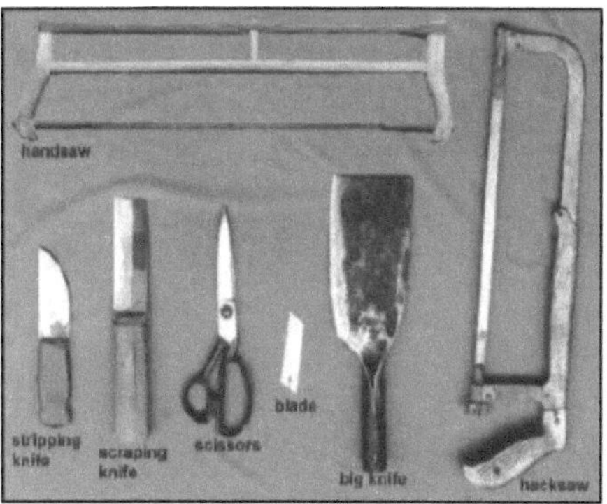

Figure 7.14: Hand Tools for Bamboo Construction (United Nations Industrial Development Organization (UNIDO), 2008)

7.7   Examples of bamboo construction in South Africa and surrounds

Bamboo in its natural form is not commonly used as a structural material in South Africa, being used mainly as fencing or decoration. While researching possible case studies in South Africa, no structures currently in use were found where bamboo was used as a primary structural material.

However, a derelict building exists on a farm outside Vredendal, where bamboo was used as the main structural elements (refer Figure 7.15 and Figure 7.16). The species *Bambusa balcooa* was brought to the area by the van Zyl family, date unknown (NBASA, 2017).

Figure 7.15: Derelict bamboo building outside Vredendal (NBASA, 2017)

Figure 7.16: Interior of a derelict building with bamboo members (NBASA, 2017)

7.8 Factor affecting construction using South African bamboo

Many factors can affect the success or failure of a building project, not all of which can be quantified. Van der Lugt (2005) defines a factor of failure or success concerning bamboo building projects as "a factor that has a negative (failure) or positive influence (success) on the costs, construction time or quality of a building project, caused by the use of bamboo, with respect to building materials more commonly used". In order to ensure the success of a building project, it is necessary to understand the limitations inherent in the chosen building material and to select the appropriate mitigations.

Structural engineering issues often found when building with bamboo could be categorised into three main areas, namely:

1) Geometry of the plants
2) Design aspects
3) Material and engineering properties

Although the following are not usually considered as structural engineering issues, these aspects should be also be taken into consideration as they might affect the design approach and construction details:

- Familiarity with and understanding of local skills with regard to bamboo construction
- Local availability of a suitable bamboo species
- Local conditions (favourable or unfavourable)

- Regulatory environment

7.8.1 Geometry

Bamboo culms do not have a regular circular shape; being closer to an ellipse. In addition, the diameter and wall thickness of the culms vary along their length, and the internodal lengths vary from base to top.

The variations in physical geometry present problems with the design and construction of bamboo structures, as discussed in Chapter 5. When building with steel, timber or concrete, the member sizes can be matched, so that connections can take the same form and aesthetic. With bamboo members, either each connection should be unique, or connectors should be used that are adaptable to varying diameters and sizes, such as those discussed in Chapter 5. In addition, the culms are hollow, which results in little or no support in the middle of the culm to transfer the loads, particularly shear loads. Thus, it is recommended that bamboo members be designed to carry mainly axial loads, with shear loads applied at the nodes.

As described in Chapter 3, *Bambusa balcooa* grows to an average height of 20 – 30 m with a culm diameter of 80 – 150 mm. The cavity is approximately one-third of the culm diameter, and the nodes are placed at 200 to 400 mm apart. Although these typical physical characteristics are known, the actual geometric dimensions are specific to each plant and would need to be determined prior to design and construction.

The curvature of the culms could affect the regularity of a structure, where the columns and beams could be slightly out of plumb. This curvature can be overcome by using a mortar overlay to mask the variations. Alternatively, the culms can be selected to minimise this curvature, and allow the slight curvature to add to the natural look of the structure.

The physical variations in the dimensions of the culms can be overcome to a certain extent by using culms harvested from commercial farmers. These culms are known to be more uniform in their dimensions, which will make the design, and in turn the construction, with such elements simpler and faster. In turn, simpler and faster designs can be assumed to be cheaper (Solomon-Ayeh, 2010).

7.8.2 Design aspects

As well as the problems associated with construction, the variations in physical geometry present problems with the design. Most structural engineering formulae and design software assume that the members being designed are constant along their length, with isotropic material properties. Not only do the culms vary along their length, but due to the composition of the fibres, they are also anisotropic, having different material and engineering properties in the longitudinal, radial and tangential directions. There are several guidelines available in the literature to overcome these issues, such as assuming average properties along the length of the culm and using actual properties when designing detailed connections. These are covered in more detail in Chapter 5.

Irregular geometry could also result in irregular structures. The irregularities can either be minimised by careful selection of the culms to be used or incorporated as part of the aesthetics of the structures, such as was done at the Bamboo Koening restaurant in Bali (refer Figure 7.17).

The bamboo culms are covered with a layer of silica on the outside. Although this waxy layer protects the culm from ingress of water, it also presents a problem during construction, as the outer layer of silica on the culm walls reduces the friction of the culm surface. This condition necessitates that the connection designs do not rely on frictional resistance of the culm to function, but rather assumes frictionless surfaces (Leake, et al., 2010).

Figure 7.17: Bamboo Koening restaurant, Bali (Nurdiah, 2016)

7.8.3 Material and engineering properties

The species *Bambusa balcooa* is reported to have a density of 720 - 850 kg/m^3 (Naik, 2005), which is similar to timber which varies from 450 - 800 kg/m^3 and is considerably lighter than steel at 7800 kg/m^3.

The lightness of this material is advantageous in conventional construction, as the low mass will reduce the load on the foundations; thus, the foundations can be smaller and less costly. In a seismic situation, the light weight will result in less downward load, resulting in a smaller seismic force on the structure. Furthermore, the flexibility and ductility of the bamboo allow the structure to withstand the forces with minimal damage.

The South African loading code for buildings, SANS 10160-4, provides requirements for buildings in seismic areas. Bamboo complies with these requirements, being lightweight. In addition, buildings in seismic areas should be of regular plan and height, which is suitable for bamboo buildings (SANS 10160-4, 2011). There are also several guidelines for the construction of seismic-resistant buildings, such as those from the Indian Plywood Industries Research and Training Institute (City University London, 2005).

However, in high wind situations, more weight in the building is preferable, to counteract the wind forces. In areas of high wind forces, heavier foundations or sturdier structures would improve the stability of the structure and assist in counteracting the wind forces. In addition, the superstructure should be constructed to function as a unit, thus increasing the resistant mass of the building.

Typically, industrial warehouses and long-span roofs are designed and constructed from structural steelwork, because concrete beams become too bulky, whereas steel can still be slender for the same properties. Conventional gang nail timber trusses can span to approximately 9 m, thereafter laminated timber beams become preferable, which increases the cost of the project.

Bamboo is lighter than steel, and trusses can be formed to span such lengths, thus making it suitable for such applications.

According to the literature available, untreated bamboo structures typically have a life span of 10 – 15 years (Janssen, 2000), being highly susceptible to attack by fungi and insects. In addition, bamboo rapidly deteriorates when exposed to moisture. However, with preservation, this can be extended to approximately 25 years, according to the literature available (Janssen, 2000). Typically, buildings are designed for a 50-year life span; thus, the shorter lifespan of bamboo could limit the usage of this material or restrict its usage to temporary or semi-permanent structures. In addition, the cost of the preservation of the bamboo culms will increase the cost of the structures built, in terms of total lifetime cost. However, this contrasts with the derelict building near Vredendal, made from bamboo members. Although the age of the building is unknown, it is presumed to be older than 50 years, and the bamboo members are still functional. Further research is required to establish expected life spans of structures, with focus on the bamboo species found in South Africa. These expected life spans would be dependent on local climates, pests and detailing of such structures.

As mentioned in Chapter 4, bamboo fails most often in shear by splitting and cracking, due to the hollow structure and lack of cross fibres in the internodes. This failure occurs particularly if the loads are applied away from the culm nodes, where the bamboo culm is weakest in shear resistance. In order to overcome this, bamboo structures should be designed to avoid internodal point loads, transferring all loads at the nodes of the culm. Alternatively, specialised steel connections may be designed to transfer the loads, with a strengthening of the nodal points of the bamboo culm.

This failure mode of bamboo presents a modicum of safety, as a failure in this way does not indicate complete failure of the entire culm, but instead indicates imminent failure. In this way, a margin of safety is available for an escape from the structure prior to collapse.

Although *Bambusa balcooa* is the best-known species in South Africa, there have been issues with the growth of this particular species in South African conditions. Attention is also being given to species such as *Dendrocalamus asper*, *Bambusa vulgaris* and *Oxytenanthera abyssinica*. The *Beema bamboo*, which is a cultivar of *Bambusa balcooa*, is also receiving much attention because of its high biomass content, which makes it more advantageous for bioenergy production.

7.9   Conclusions

It can be seen that bamboo is available in South Africa, potentially for use as a structural material. Although the quantity of currently available bamboo for structural purposes has not been established, South Africa has the potential to expand its production thereof. As the demand for bamboo increases, so too, the supply would increase.

The engineering and material properties of *Bambusa balcooa* found in South Africa have not yet been established. Testing would be required to determine these design parameters before bamboo construction could become commonplace in South Africa. These tests would also assist in obtaining building permits in South Africa, as well as the creation of a South African building code for bamboo.

Furthermore, investigation should be done into the properties of *Beema bamboo*, as an alternative species to *Bambusa balcooa*.

Once the tests have been done, and the building permits received, it follows that the local labour can be instructed in the necessary skills for bamboo construction during the construction of a bamboo building. So too, the number of bamboo buildings in South Africa would increase, which would encourage the broader use of bamboo as a building material.

The limitations for bamboo design and construction are not unique to South Africa but are common to other countries involved in bamboo construction. Their experience in overcoming these limitations can be transferred to the use of bamboo in South Africa, making bamboo construction a viable building technology in South Africa.

7.10   References

Anon., 2017. *Bathurst bamboo plantation grows a sustainable future.* [Online] Available at: www.talkofthetown.co.za/2017/11/29/bathurst-bamboo-plantation-grows-sustainable-future/ [Accessed 15 November 2019].

Bamboo Interim Steering Committee, 2012. *Business Case for the National Bamboo Association of South Africa (NBASA)*, Pretoria: s.n.

Barathi, N., 2019. *Beema Species.* [Online] Available at: www.growmorebiotech.com/beema-bamboo.html [Accessed 30 November 2019].

Bystriakova, N., Kapos, V. & Lysenko, I., 2002. *Potential distribution of woody bamboos in Africa and America,* Beijing: INBAR.

Bystriakova, N., Kapos, V. & Lysenko, I., 2004. *Bamboo biodiversity: Africa, Madagascar and the Americas.* [Online] Available at: http://www.unep-wcmc.org/resources/publications/UNEP_WCMC_bio_series/19.htm [Accessed 10 09 2011].

Canavan, S., Richardson, D., Le Roux, J. & Wilson, J., 2018. Alien Bamboos in South Africa: A Socio-Historical Perspective. *Human Ecology,* 29 November, 47(1), pp. 121-133.

Chung, K. & Chan, S., 2002. *Design of Bamboo Scaffolds,* Beijing, China: INBAR.

City University London, 2005. [Online] Available at: http://www.staf.city.ac.uk/earthquakes/Bamboo/Bamboo.htm [Accessed 03 06 2011].

Food & Trees for Africa, 2011. *Bamboo for Africa.* s.l.: s.n.

INBAR, 2000. *Technical report 17: Bambusa balcooa.* [Online] Available at: http://www.inbar.int/publication/txt/tr17/bambusa/balcooa.htm [Accessed 27 02 2012].

INBAR, 2015. *Bamboo for Africa: A strategic resource to drive the continent's Green Economy,* Beijing, China: INBAR.

INBAR, 2019. *Trade Overview 2017 - Bamboo and Rattan Commodities in the International Market,* Beijing: INBAR.

INBAR, 2020. *Trade Database,* Beijing: INBAR.

Janssen, J., 2000. *Designing and building with bamboo,* Beijing: INBAR.

Kaminski, S., Lawrence, A. & Trujillo, D., 2016. *Design Guide for Engineered Bahareque Housing,* Beijing: INBAR.

Kibwage, J. K., Frith, O. B. & Paudel, S. K., 2011. *Bamboo as a building material for meeting East Africa's housing needs: a value chain case study from Ethiopia,* Beijing: INBAR.

Leake, G., Toole, K., Divis, P. & Torres-Sanchez, C., 2010. *Bamboo as a solution for low-cost housing and storage in Pabal (India).* s.l., University of Strathclyde.

Lekhetho, S. & Notten, A., 2019. *Bergbambos tessellata [image].* [Online] Available at: www.pza.sanbi.org/bergbambo-tessellata [Accessed 15 November 2019].

Naik, N. K., 2005. *Mechanical and Physico-Chemical Properties of Bamboo,* Mumbai: Indian Institute of Technology.

NBASA, 2017. *Bamboo in Vredendal,* Vredendal: NBASA.

Nurdiah, E. A., 2016. The Potential of Bamboo as Building Material in Organic Shaped Buildings. *Procedia - Social and Behavioral Sciences,* Volume 216, pp. 30-38.

Paudel, S. K. & Ayeh, S., 2003. *Bamboo School Building,* Beijing: INBAR.

Richard, M. J., 2013. *Assessing the Performance of Bamboo Structural Components,* Pittsburgh: University of Pittsburgh.

S. Ross, 2017. *Bambusa balcooa.* Cape Town: Private collection.

SANS 10160-4, 2011. *Basis of structural design and actions for buildings and industrial structures - Part 4: Seismic actions and general requirements for buildings.* Pretoria: SABS.

SANS 10400-K, 2015. *The application of the National Building Regulations Part K: Walls.* 3.1 ed. Pretoria: SABS.

SANS 10400-XA, 2011. *The application of the National Building Regulations - Part XA: Energy usage in buildings.* Pretoria: SABS.

Scheba, A., Blanchard, R. & Mayeki, S., 2017. *Bamboo for green development? The opportunities and challenges of commercialising bamboo in South Africa,* Cape Town: HSRC.

Solomon-Ayeh, K. A., 2010. *Use of bamboo for buildings - a sustainable, strong, versatile and economic option for the preservation of timber in Ghana,* Kumasi, Ghana: Building and Road Research Institute (BRRI).

Spears, C., 2008. *Yushania alpina.* [Online] Available at: http://www.en.wikipedia.org/wiki/Yushania_alpina [Accessed 15 February 2018].

Stamm, J., Tesfaye, M. & Girma, H., 2014. *Construction Manual with Bamboo,* Addis Ababa, Ethiopia: African Bamboo.

United Nations Industrial Development Organization (UNIDO), 2008. *Cottage Industry Manuals: Raw Materials and Tools for Bamboo Applications, Eastern Africa Bamboo Project,* Amsterdam: INBAR.

van Asch, M., 2009. *Bamboo documentation,* Dassenberg: Alba Laboratories.

van Rijswijck, E., 2011. *Bamboo shoots up in Eastern Cape.* [Online] Available at: http://www.mediaclubsouthafrica.com/economy/2698-bamboo-051211 [Accessed 24 April 2013].

Victor, J. E., 2006. *Distribution of Thamnocalamus tessellatus [image].* [Online] Available at: http://redlist.sanbi.org/imgs/distribmaps/9/Plants-93257.png [Accessed 25 February 2018].

Chapter 8

8   EVALUATION, CONCLUSIONS AND RECOMMENDATIONS

8.1   Summary and Evaluation

Bamboo has been used as a traditional construction material for centuries to build structures such as bridges, houses, as well as warehouses and similar buildings. More recently, it has been presented as a preferred material from a sustainability and ecological impact perspective. Although bamboo construction is an accepted form of construction in some countries, little is known about this type of construction in South Africa.

This book has researched and collated available information regarding various aspects of bamboo construction, from a material and structural engineering perspective, focussing on the use of whole culms as structural elements. In this process, the following objectives of this book were fulfilled:

1)   The material and engineering properties of eleven species of bamboo were collated and evaluated. These species were *Bambusa balcooa, Bambusa nutans, Bambusa polymorpha, Bambusa stenostachya, Bambusa vulgaris, Dendrocalamus asper, Dendrocalamus giganteus, Dendrocalamus strictus, Gigantochloa apus, Guadua angustifolia* and *Phyllostachys pubescens.* Information was gathered from online news articles, project reports and internet websites. Structural journals were also reviewed for articles on bamboo construction.

Each document or article retrieved was critiqued for the reliability and validity of the information that it presented, by comparison with similar articles. In addition, articles or documents that presented statements regarding the properties of bamboo plants without numerical statistics were discarded.

2)   The existing design codes, standards and guidelines for full culm bamboo construction were assessed. Furthermore, various proposed design approaches were presented and critiqued, including the transfer of these design approaches to computer modelling and design. Designing for seismic situations was also covered briefly, as bamboo construction is considered a favourable building material for such situations due to its light weight and high flexibility.

3)   Four case studies were selected from relevant global examples to illustrate the viability and possibilities for the use of full-culm bamboo in buildings and structures. These were a pedestrian bridge spanning 52 m, a warehouse with trusses of 33 m in length, a residential house made entirely from bamboo culms and bamboo products, and bamboo pavilions designed for a community centre.

4)   One naturalised species of bamboo found in South Africa, *Bambusa Balcooa,* may be considered suitable for whole culm bamboo construction. However, the particular engineering and material

properties of this species in South Africa have not yet been established. Considerable testing and proving would be required before this could be considered a viable building material. Other imported species such as *Phyllostachys edulis*, *Bambusa vulgaris* and *Oxytenanthera abyssinica*, are currently being grown and assessed to determine which ones adapt best to the climate in various parts of South African.

Several key questions regarding building with bamboo in its natural form were raised in Chapter 1; these questions will be addressed here.

8.1.1   Can the building and construction sector reduce Greenhouse Gas emissions?

In 2018, 11% of global greenhouse gas (GHG) emissions were attributed to the construction industry, 23% to the transport industry, and 28% to emissions from buildings during their lifetime (IEA, 2019).

The transport sector included emissions from road transport, as well as aviation, rail and shipping, while the building sector included combustion of biomass and fossil fuels in residential and commercial buildings, mostly for cooking and heating.

It was found that reports produced by different bodies or groups provided varying numbers regarding GHG emissions. The percentages presented here were extracted from global reports that focussed specifically on the building and construction industry, as these reports were deemed to have the most accurate figures for this topic.

A separate report indicated that in 2018 in South Africa, 16% of the national GHG emissions were attributed to the construction industry. However, this figure included the production of products such as cement and steel.

As these figures were extracted from separate reports, it is unclear whether the same parameters were used to determine the percentages in each report or not. If they were, this would indicate that the South African construction industry is responsible for a higher percentage of GHG emissions than the global average. By following global trends in material selection in buildings, it should be possible to reduce the GHG emissions in the South Africa construction industry.

One of the methods used to determine the GHG emissions of a material through its life cycle is the measurement of its carbon footprint, sometimes also referred to as the embodied carbon of a material or product. By selecting a construction material with a low carbon footprint, which considers the transport of the material from its site of manufacture to the construction site, the construction industry can reduce or minimise these emissions. Outside of the United Kingdom and in the absence of country-specific data, the ICE database (ICE, 2019) can be used as 'proxy data' to determine the embodied carbon of a chosen material.

In addition to selecting a material with a low embodied carbon, by selecting materials that are available close to the site of construction, transport emissions could be minimised. Lastly, the selection of building material affects the energy required by buildings during their lifetime for heating and cooling purposes, and the careful selection of such would also affect these emissions.

8.1.2   What is bamboo and where is it found?

Bamboo is a member of the grass family, *Poaceae,* and the subfamily *Bambusoideae.* There are more than 1600 known species of bamboo, although not all of these are suitable for use as construction elements. Eleven species commonly used in construction were identified and researched in this book, as follows:

- *Bambusa balcooa*
- *Bambusa nutans*
- *Bambusa polymorpha*
- *Bambusa stenostachya*
- *Bambusa vulgaris*
- *Dendrocalamus asper*
- *Dendrocalamus giganteus*
- *Dendrocalamus strictus*
- *Gigantochloa apus*
- *Guadua angustifolia*
- *Phyllostachys pubescens*

The bamboo plant is made up of a visible stem, known as the culm, leaves, and an underground root system. The culm is the portion of the plant used in building construction and is usually hollow, although certain species have solid culms. Bamboo plants adapt to their differing natural environments, and so vary in diameter, length and colour, depending on the local climate. Thus, the macro- and microstructure of the bamboo culms vary between species, as well as their location.

Bamboo plants are found mainly in the tropical, subtropical and mild-temperate regions of the world. South Africa has one indigenous species of bamboo, *Thamnocalamus tessellatus*, which is used to construct gates and screens, and one naturalised species, *Bambusa balcooa*, which is used as a structural material in other parts of the world.

8.1.3   Can bamboo be considered a sustainable construction material?

The terms 'sustainable construction' and 'green building' are often used interchangeably to define construction methods that apply some of the principles of Sustainable Development. Kibert defines sustainable construction as "creating a healthy built environment based on ecologically sound principles" (Kibert, et al., 2002), while the National Association of Home Builders in the United States of America defines green building as building in a manner that conserves resources (NAHB, 2002).

Two aspects of sustainable construction that apply to bamboo construction are resource efficiency and the use of renewable materials. The aim is to reduce the environmental impact of the building construction by the selection of suitable building materials, specifically those with a low carbon footprint of embodied carbon (refer discussion under item 8.1.1).

Although many publications state that bamboo is a sustainable building material, comparing favourably with other building materials, the quantitative proof appears to be lacking or unclear. Different methods have been used to evaluate bamboo, which makes comparison and evaluation of the reports difficult.

Using Life Cycle Assessment (LCA) and Eco-indicators as a basis for comparison, Murphy (2004) compared a conventional masonry house and a house of similar size constructed using *Guadua angustifolia* in the bahareque method (bamboo structural frame with cement-plastered walls). He concluded that the bamboo house had approximately 60% of the environmental impact of the conventional masonry house (Murphy, et al., 2004). Murphy further indicated that 95% of the environmental impact for the bamboo house came from the foundations and the wall plaster, both of which contained cement.

Van der Lugt (2003) also compared the environmental impacts of components in a bamboo pedestrian bridge made from *Guadua angustifolia* bamboo, with timber, steel, and concrete elements using Life Cycle Assessment (LCA), but with the environmental impacts of each converted into an environmental cost for comparison purposes. He concluded that in these applications, bamboo could be considered the most sustainable alternative for all elements, attributing this result to the structurally efficient cross-section of the bamboo culm, and the simple production processes involved (Van der Lugt, et al., 2003). However, after carrying out a life cycle costing (LCC) analysis, he concluded that steel was a more cost-effective material than bamboo, due to bamboo's shorter life span and higher labour costs associated with assembling, based on the local labour rates and experience available at that time (Van der Lugt, et al., 2003).

Although both these studies used LCA as a basis to compare bamboo with conventional building materials, Murphy compared a building in its entirety while Van der Lugt considered the components of the bridge separately. Although both approaches have merit, it is more common to select materials for a particular component of a structure, such as a beam, as certain materials might be better suited for particular functions. In addition, Van der Lugt included the lifespan of bamboo in his assessment and concluded that steel was a more cost-effective solution in that instance due to the short lifespan of the bamboo.

Although bamboo can be considered a favourable choice of construction material when looking at certain environmental aspects, this does not automatically mean it is the best choice. All aspects affecting the environmental impact should be considered, using a cradle-to-grave approach. Further research would be required into local species and costs to determine if bamboo construction is an ideal choice for local conditions.

8.1.4 What are the engineering and material properties of bamboo?

Each bamboo species has its own particular anatomical structure, which determines the material and engineering properties thereof. The key anatomical characteristics affecting these properties can be summarised as follows:

- Variation in the number of fibres in the culm wall
- Variation in the cross-section of the culm along its length
- Moisture content of the culm
- Age of the culm
- Environmental growth conditions

The engineering and material properties of the eleven species, as mentioned in Section 8.1.2, were discussed in Chapter 4. A full summary of the physical and mechanical properties, as well as other documented results, are given in Appendix A. These properties included density, Elastic Modulus, Poisson's Ratio, shrinkage, creep, thermal performance, axial compression and tension, flexural strength and shear.

It was found that the properties documented in the literature for the various species varied to a large extent. In many instances, the exact testing procedures were not documented. As a result, it was unclear whether the discrepancies within the same species were due to the testing procedures or due to factors such as environmental growth conditions. As bamboo is a natural material, i.e. not man-made, it was presumed that the variances in properties were due to local growing conditions. These variances highlighted the need for testing of the particular species used in the design process before construction thereof.

8.1.5 What are the existing standards and design codes for bamboo construction?

A design code can be defined as "a set of rules that specify the minimum standards for constructed objects such as buildings" (Anon., 2016), while a design standard can be defined as "Technical specifications or rules, based on consensus, and approved by a recognised standardising body" (Janssen, 2000).

There are various design codes and standards available for bamboo design and construction as detailed in Chapter 5, for the countries where bamboo plants grow in abundance, such as China, Colombia and India. There is also an international design standard for bamboo construction, namely ISO 22156. Some codes, such as the National Building Code of India, are available in English, while others require translation, such as the Colombian design code. The language barrier resulted in limitations in comparing the different codes and standards, although several the illustrations appeared similar.

For countries where there is currently no bamboo design code or standard, such as South Africa, it is recommended that the International Standard, namely ISO 22156: Bamboo-Structural design be used as a design basis, using specific material properties relevant to the local species. This standard follows a Limit States Design approach, with the focus on the performance of the bamboo structure. Design stresses are not provided, but the designer is instructed to use strength and stiffness parameters derived from local tests.

In addition, there are guidelines as published by INBAR which can be used for the preliminary sizing and design of such structures, such as *Designing and Building with Bamboo – Technical Report 20* (Janssen, 2000) and *Design of Bamboo Scaffolds – Technical Report 23* (Chung & Chan, 2002). There are also several construction guidelines available, such as the *Bamboo Construction Source Book* by the Hunnarshala Foundation (2013) and the *Construction Manual with Bamboo* by Stamm et al. (2014), which can be used as a starting point for the bamboo structures. However, the details would still require design checks.

There appears to be sufficient information pertaining to recommended design guidelines for bamboo structures, provided that the material and engineering properties of the particular bamboo species to be used have been determined. As bamboo construction becomes a more commonplace type of construction, it can be expected that more countries would incorporate the design of bamboo into their local building codes and standards.

8.1.6   What design procedures would be used to design a bamboo structure?

For countries without bamboo design codes or standards, such as South Africa, the international ISO codes 22156 and 22157 could be used for design purposes, with the appropriate material properties obtained from the bamboo species considered. In addition, the calculation methods and approaches in the design codes from India, Peru, Ecuador and Colombia could be used as a guideline.

Where no reliable material properties of a specific bamboo species are available, guidelines are available for preliminary sizing of the various structural elements and members. Once the properties of the species to be used have been determined, these values can be inserted into the calculations for optimisation of the designs. Based on the particular structural characteristics of bamboo culms, several key characteristics should be borne in mind when designing a bamboo structure, as follows:

- Bamboo culms are weak in both compressive and tensile resistance when loaded perpendicular to the culm, i.e. shear capacity is low.
- Bamboo is stronger, and connections are simpler to construct, when the loads are parallel to the fibres of the culm, i.e. along the longitudinal axis of the culm.

- It is difficult to use the full tensile capacity of the culms, as the tensile strength of the connections governs the maximum tensile capacity.

- Bamboo is more flexible and bends more than a timber member of similar outside dimensions.

- Bamboo culms are often curved, so buckling of slender sections in compression should be taken into consideration.

- If untreated and exposed to the environment, bamboo is prone to rot and insect attack (Kaminski, et al., 2016).

Although bamboo structures can be modelled on computer design software, the non-homogenous nature of the material creates difficulties as these design approaches assume that a design element consists of a uniform material throughout. Depending on the complexity of the structure as well as the level of analysis required, two approaches are possible. The first approach, as suggested by Liang and Robinson (2009) and Sharma (2010), would be to model each culm as a constant, uniform hollow cylinder, with material properties derived from research, and mechanical properties such as the moment of inertia calculated from average diameters.

A second approach, as suggested by Silva (2008), would be to model a graded variation of the member properties in a Finite Element software program. He suggested that the extra effort required to model the graded variation would be justified when designing for local stresses near supports, pin connections or holes. However, the effort involved to design to this level of detail would have to be weighed up against the usefulness of the information.

Although there are several codes and standards for bamboo design, the National Building Code of India was found to be the simplest to follow. However, no formulas were provided; instead, the designer is recommended to design using the principles of beam theory, for instance.

In contrast, the Colombian design code provides element and connection design methods for bamboo construction, using an allowable stress design approach. Use of this code would necessitate translation into English, but it might provide valuable guidance which does not appear to be available in the other codes.

8.1.7 Are there examples of engineered bamboo structures?

As well as the traditional bamboo structures found globally, there are many instances of engineered bamboo structures. Four case studies from different construction sectors were reviewed to demonstrate the wide range of applications.

The first two case studies illustrate the use of bamboo construction in scenarios where the common choice of materials would be structural steelwork. The first describes the design and construction of a 52 m span bamboo pedestrian bridge over a river in Colombia. This bridge was constructed from the species *Guadua angustifolia* with a reported compressive strength of 70 MPa and tensile strength of 200 MPa.

This case study demonstrated that bamboo could be a suitable construction material for a large structure when combined with the applicable engineering design. It also illustrated the problems that could be encountered if the design intent is not sufficiently transferred to those involved in the construction.

Although the cost of this bridge was reported to be similar to the cost of a structural steel bridge, the designer noted that this span of 52 m was close to the economical limits for bamboo arch bridges. This bridge was also designed for pedestrian traffic only. This designer reported only one bamboo bridge designed to carry vehicle loading, and that was designed to carry a single 2-ton vehicle at a time. It is unclear whether bamboo would be suitable or economical for trafficable bridges.

The second case study describes the use of bamboo trusses in a storage warehouse in Bali. These trusses were constructed from two species, namely *Dendrocalamus asper* and *Gigantochloa apus*, all chosen to be approximately 140 mm in diameter, although no strengths were given for the actual bamboo used. No engineering design details were provided, but from the photographs, it appears that the frame made use of 16 m long Vierendeel trusses in a portal frame system.

Although this warehouse was only 16 m in width, the prototype design was for spans up to 20 m. Considering the light weight of bamboo when compared with similar steel round sections, this would appear to be a viable alternative to conventional steel trusses and frames, for buildings within this width range.

The third case study represents the use of bamboo construction in a residential scenario. It describes a house built entirely from bamboo, which was an entry in the Solar Decathlon Europe, held in Spain in 2010. Although the focus of the decathlon was the production of self-sufficient buildings regarding energy use, it also clearly demonstrates the use of bamboo for all structural components of a single storey building. Although no mention was made of the bamboo species used for the Solar House case study, it was assumed that this was constructed from *Phyllostachys pubescens*, one of the most common species in China. The compressive and tensile strengths as reported in Chapter 4 for this species are 56 MPa and 196 MPa, respectively.

No engineering information was available regarding the design or construction of this house. As a result, it is inconclusive from this case study as to whether bamboo is suitable for such an application.

The fourth case study represents the use of bamboo construction in the commercial sector. It expands on the use of bamboo for all building components, describing a community centre in Vietnam which uses local materials and traditional bamboo construction skills to create a modern architectural concept. Although no mention was made of the bamboo species used, it was assumed that this was constructed from the species found in that location, namely *Bambusa stenostachya*, with a compressive strength of

51 - 66 MPa and tensile strength of 8 MPa. No information was found regarding structural engineering input; however, the architectural firm Vo Trong Nghia has in-house engineers, and it can be presumed that the engineers had input into these designs.

Although engineering information on bamboo structures is limited or unavailable, from these case studies, it has been shown that such structures have been built, and that bamboo construction is not limited to small structures.

8.1.8 Are there any limitations to designing with bamboo?

Although bamboo is considered a favourable construction material from many viewpoints, there are some limitations that need to be considered when designing and constructing with this material, as discussed in the preceding chapters.

Bamboo culms vary physically in diameter, intermodal length, wall thickness and composition, and structurally regarding material and structural properties. These variations occur between species, but can also occur within the same species, for plants grown in different locations.

Bamboo culms do not have a regular circular shape; being closer to an ellipse. The diameter and wall thickness of the culms vary along their length, and the culms often curve slightly along their length as well.

The variations in physical geometry present problems with the construction of bamboo structures. When building with steel, timber or concrete, the member sizes can be matched, so that connections can take the same form and aesthetic. With bamboo members, each connection must either be unique or must use connectors that are adaptable to varying diameters and sizes, such as those discussed in Chapter 5. In addition, the culms are hollow, which results in little or no support in the middle of the culm to transfer the loads, particularly shear loads. Thus, it is recommended that bamboo members be designed to carry mainly axial loads, with shear loads applied at the nodes.

As well as the problems associated with construction, the variations in physical geometry present problems affecting the design. Most structural engineering formulae and design software assume that the members being designed are constant along their length, with isotropic material properties. Not only do the culms vary along their length, but due to the composition of the fibres, they are also anisotropic, having different material and engineering properties in the longitudinal, radial and tangential directions. There are several guidelines available in the literature to overcome these issues, such as assuming average properties along the length of the culm and using actual properties when designing detailed connections.

Untreated bamboo structures typically have a life span of 10 – 15 years (Janssen, 2000), being highly susceptible to attack by fungi and insects. However, with preservation, this can be extended to 25 years (Janssen, 2000). Typically buildings are designed for a 50-year life span; thus, the shorter lifespan of bamboo could limit the usage of this material. In addition, the cost of the preservation of the bamboo culms will increase the cost of the structures built, in terms of total lifetime cost, as well as increasing the environmental impact of such structures.

8.1.9   Is bamboo construction a viable building technology for South Africa?

Although there are possibilities to use other species of bamboo for construction, there are two species in South Africa currently considered suitable, namely *Bambusa balcooa*, a naturalised species, and *Beema bamboo*, a clone of *Bambusa balcooa*. Unfortunately, some of the attempts to grow these species of bamboo have failed, and the remaining growers and nurseries do not have quantitative production figures. However, local reports indicate that South Africa has the potential to expand its bamboo growth industry to meet potential demand.

The engineering and material properties of *Bambusa balcooa* as found in South Africa have not yet been established. Testing would be required to determine these design parameters before bamboo design and construction could become commonplace in South Africa.

As bamboo design is not included in any South African design code, it would be necessary to use codes and guidelines from other countries, as described under Section 8.1.5, such as the International Standard ISO 22156: Bamboo-Structural Design or *Designing and Building with Bamboo* as published by INBAR.

The limitations for bamboo design and construction are not unique to South Africa but are common to other countries involved in bamboo construction. Their experience in overcoming these limitations can be transferred to the use of bamboo in South Africa, making bamboo construction a viable building technology in South Africa.

8.2   Conclusions

The preliminary literature review indicated that there was a lack of scientific or structural information regarding bamboo as a building material from an engineering or materials perspective, both globally as well as in South Africa.

Although a substantial amount of information was subsequently found on the various aspects of bamboo as a structural material, the information for the eleven species investigated varied widely between sources. The methods used to determine the various properties were often unclear or unspecified, and the results were often presented either as single values or as an average, without providing information regarding the range of the results. In 2012 the International Standards Organisation issued their latest standard for the determination of the properties of bamboo, namely ISO 22157. Thus, it can be presumed that the results of testing for various bamboo properties would now be more consistent for the same species, as the same testing procedures would be used.

From an environmental and sustainability perspective, bamboo can be considered a favourable choice of construction material. However, the environmental assessment of bamboo is affected by factors such as transport of the bamboo to the site, which could result in bamboo not being the best alternative in a particular situation. All aspects affecting the environmental impact should be considered, using a cradle-to-grave approach. Further research would be required into local species and costs to determine if bamboo construction is an ideal choice for local conditions.

In conclusion, bamboo construction is a viable construction technology for South Africa. However, before this can become an accepted construction technology in South Africa, further research would be required. Possible areas to be considered are detailed in the following section.

8.3   Recommendations for further research

Based on the findings of this book, the following recommendations are given for future research:

8.3.1   Determination of properties of South African bamboo species

As described in Chapter 3 and 4, the physical, chemical, material and strength properties vary between different species of bamboo. Therefore, the specific properties of the bamboo to be used should be established before it is used for design and construction.

Although South Africa has a naturalised bamboo species that is theoretically suitable for construction purposes, namely *Bambusa balcooa*, the engineering and material properties of the plants grown in South Africa have not yet been established. Testing would be required to determine these design parameters to confirm the suitability of this species for construction purposes.

Furthermore, an investigation should be done into the properties of *Beema bamboo*, as an alternative species to *Bambusa balcooa*. Although this is being grown in South Africa to provide biomass feedstock, its similarity to *Bambusa balcooa* would suggest that this could also be used as a structural material.

8.3.2   Construction of bamboo test structure

Once the various properties of the South African bamboo species have been established, it would be recommended that these values are used in a design and construction of a small building. This process would highlight any construction issues pertinent to this particular bamboo species, as well as verifying the theoretical values used in the design.

8.3.3   Thermal performance of bamboo species

As mentioned in Chapter 4, very little information is available regarding the thermal properties of any bamboo species. From a sustainable construction perspective, it is important to know these properties, as they dictate the energy that will be required to heat or cool a building after construction.

This area requires further research, both into the properties as well as the possible mitigation measures required to be incorporated into the designs as a result.

8.3.4 Behaviour of bamboo in fire

Although limited information was found in the literature regarding bamboo's resistance to fire, from the research available, it would appear that bamboo has an initial resistance to fire similar to timber. However, once charring begins, its structural strength will be compromised sooner since it has thin walls and is not a solid element.

Although there are measures available to increase the fire resistance of the bamboo culms, insufficient information was found in the literature to validate the claims that treated bamboo is an acceptable building material regarding fire protection. This area requires further research if bamboo is to be considered a viable structural building material.

8.3.5 Bamboo Construction in South Africa

Chapter 7 provided an overview of bamboo construction in the South African context. An in-depth review of local conditions affecting this technology would be recommended, with specific focus on areas such as the effect of sub-regional climate on the growth of the various bamboo species, access to technology and familiarity with and understanding of local skills with regard to this type of construction

8.4 References

Anon., 2016. *Building Code.* [Online] Available at: http://en.wikipedia.org/wiki/Building_code [Accessed 20 January 2016].

Chung, K. & Chan, S., 2002. *Design of Bamboo Scaffolds,* Beijing, China: INBAR.

ICE, 2019. *Embodied Carbon - The ICE Database.* [Online] Available at: http://www.circularecology.com/ice-database.html

IEA, 2019. *2019 Global Status Report for Buildings and Construction: Towards a zero-emission, efficient and resilient buildings and cosntruction sector,* s.l.: Global Alliance for Buildings and Construction.

Janssen, J., 2000. *Designing and building with bamboo,* Beijing: INBAR.

Jayanetti, L. & Follett, P., 1998. *Technical Report 16: Bamboo in Construction - An Introduction,* Beijing: INBAR.

Kaminski, S., Lawrence, A. & Trujillo, D., 2016. *Design Guide for Engineered Bahareque Housing,* Beijing: INBAR.

Kibert, C. J., Sendzimir, J. & Guy, G. B., 2002. *Construction Ecology - Nature as the basis for green buildings.* London: Spoon Press.

Liang, C. B. & Robinson, P., 2009. *Bamboo as a Permanent Structural Component,* London: Imperial College London.

Murphy, R. J., Trujillo, D. & Londono, X., 2004. *Life Cycle Assessment (LCA) of a GuaduaHouse.* Pereira, Colombia, s.n., pp. 235-244.

NAHB, 2002. *Building Gree, Building Better: The Quiet Revolution,* Washington, D.C.: National Association of Home Builders.

Richard, M. J., 2013. *Assessing the Performance of Bamboo Structural Components,* Pittsburgh: University of Pittsburgh.

Sharma, B., 2010. *Seismic Performance of Bamboo Strructures,* Pittsburgh: University of Pittsburgh.

Silva, E. C. N., Walters, M. C. & Paulino, G. H., 2008. *Modeling Bamboo as a Functionally Graded Material,* s.l.: American Institute of Physics.

Van der Lugt, P., Van Den Dobbelsteen, A. & Abrahams, R., 2003. Bamboo as a building material alternative for Western Europe? A study of the environmental performance, costs and bottlenecks of the use of bamboo (products) in Western Europe. *Journal of Bamboo and Rattan,* 2(3), pp. 205-223.